LIFTING THE FOG OF PEACE

Lifting the Fog of Peace

HOW AMERICANS LEARNED TO FIGHT MODERN WAR

Janine Davidson

The University of Michigan Press • *Ann Arbor*

First paperback edition 2011
Copyright © 2010 by Janine Davidson
All rights reserved
Published in the United States of America by
The University of Michigan Press
Printed and bound by CPI Group (UK) Ltd, Croydon, CR0 4YY

2014   2013   2012   2011   5   4   3   2

No part of this publication may be reproduced, stored
in a retrieval system, or transmitted in any form or by
any means, electronic, mechanical, or otherwise,
without the written permission of the publisher.

A CIP catalog record for this book is available from the British Library.

Library of Congress Cataloging-in-Publication Data

Davidson, Janine.
    Lifting the fog of peace : how Americans learned to fight modern
war / Janine Davidson.
      p.   cm.
    Includes bibliographical references and index.
    ISBN 978-0-472-11735-2 (cloth : alk. paper)
    ISBN 978-0-472-02298-4 (ebook)
    1. United States—Military policy.   2. United States—Armed
Forces.   3. Military doctrine—United States.   4. Counterinsurgency—
United States.   5. Nation-building—United States.   I. Title.

UA23.D2753   2010
355'.033573—dc22   2010007916

ISBN 978-0-472-03482-6 (pbk. : alk. paper)

The views presented in this book are those of the author and do not necessarily
represent the views of the Department of Defense or its Components.

Contents

Acknowledgments vii

List of Acronyms xi

Introduction: On the Front Lines with America's Nation Builders   1

CHAPTER 1. Military Learning and Competing Theories of Change   9

CHAPTER 2. Two Centuries of Small Wars and Nation Building   27

CHAPTER 3. Vietnam to Iraq: Debating the "New World Order"   67

CHAPTER 4. Learning to Learn: The Training Revolution in the Post-Vietnam Military   97

CHAPTER 5. Doctrine and Education for the New Force   129

CHAPTER 6. Learning to Surge in Iraq   159

Conclusion: Learning Theory and Military Change in the 21st Century   191

Appendix: Key Terms and Conceptual Confusion   203

Bibliography   209

Index   227

Acknowledgments

This book would not have been possible without the support and assistance of a number of remarkable people and institutions. First, my graduate school adviser at the University of South Carolina, Dr. Jerel Rosati, was a great source of personal and professional inspiration. From encouraging my initial pursuit of the topic in a graduate seminar, to the countless hours of brainstorming during the research and writing phases, Jerel kept me focused and optimistic. Thanks likewise to Dr. Don Fowler for constantly reminding me to "get it done!" and to Dr. Paula L'Ecuyer, who spent hours over coffee helping me clarify my thinking. Other South Carolina scholars, including Donald Puchala, Charles Kegley, and Kendrick Clements, provided a diverse and supportive team of critics and mentors, while Sally Buice kept us all on track and focused on having a bit of fun!

The bulk of this research was accomplished while I was in residence as a research fellow at the Brookings Institution in Washington, DC. I am deeply indebted to Michael O'Hanlon and Jim Steinberg at the Brookings Foreign Policy program for generous support and for the opportunity to work in such an intellectually vibrant environment. Brookings Army Fellow Colonel David Gray provided the unique and inspiring perspective of a decorated soldier who is also a distinguished military historian. By introducing me to the "real Army" and constantly challenging me to think beyond conventional wisdom, Colonel Gray was a true mentor and friend. Special thanks are also in order to Brookings scholar Peter Singer and the 21st Century Defense Initiative, who provided me additional support as a nonresident fellow while I finished the book in 2008.

I am also indebted to a number of distinguished members of the national security community who were willing to take time out of their busy schedules to meet with me during the research phase of this project.

Among these, thanks are in order for Eliot Cohen, Michèle Flournoy, Johanna Mendelson Forman, Jim Mattis, Jim Miller, Sarah Sewall, and Anthony Zinni. I owe special thanks to Joe Collins, David Fastabend, Bob Killebrew, and Greg Newbold, each of whom either read and reread chapter proofs or provided valuable insight, encouragement, and personal mentorship throughout the different phases of this project.

I must also thank another very special network of scholars and practitioners, including John Agoglia, Beth Cole, Conrad Crane, Johanna Mendelson Forman, TX Hammes, Frank Hoffman, Colin Kahl, Elisabeth Kvitashvili, Richard Lacquement, Montgomery McFate, Kathleen McInnis, John Nagl, Linda Robinson, Nadia Schadlow, Tammy Schultz, Nina Serafino, Erin Simpson, Vikram Singh, and Mark Smith, all of whom provided valuable insight, constructive commentary, and good humor through hours of informal discussion and debate on the topics of counterinsurgency and stability operations. Much of the insight on current military issues was gleaned through workshops and seminars sponsored by the Army's Peacekeeping and Stability Operations Institute (PKSOI), the Philip Merrill Center for Strategic Studies at the Johns Hopkins School for Advanced International Studies (SAIS), the Center for New American Security (CNAS), Hicks and Associates (H&AI), Women in International Security (WIIS), and the U.S. Institute of Peace (USIP). David Dilegge and Bill Nagle, founders and tireless editors of the *Small Wars Journal* website and blog, deserve a special thanks for providing a creative and free forum for this community's debate and thought leadership.

This research was also greatly enhanced by the insight provided by a number of professional civilians, military officers, and enlisted troops who assisted me in site visits and interviews. I am grateful to the soldiers of the First Brigade (Bastogne!), 101st Airborne Division; the instructors at the First Marine Division Infantry School; the Observer Controllers at the Joint Readiness Training Center; the military staff of the First Regiment, Second Marines; the instructors and professors at the Army Command and General Staff College, the Army War College, and the Marine Corps Command and Staff College; the analysts at the Center for Army Lessons Learned and the Marine Corps Warfighting Laboratory; the staff of the Marine Corps University Archives and the Army Peacekeeping and Stability Operations Institute; and the Army and Marine Corps doctrine writers. The members of each of these organizations were incredibly generous with their time and amazingly patient with my endless questions—and follow-up questions.

I am especially indebted to my parents, Jim and Joanne, whose support was a continued source of strength; and to my sister Jennifer, whose wisdom and professional example are an inspiration. Thanks also to my brother and sister-in-law, Jim and Melanie, and my friends Jo and Ron Carlberg, who generously opened their vacation homes to me for use as a writing retreat. Large portions of this book were written in these quiet sanctuaries. Thanks also to Jennifer Sklarew at George Mason University, who assisted with editing, and Melody Herr, Scott Griffith, and Kevin Rennells at the University of Michigan Press for their patience and dedication to the project from start to finish. Finally, this book might never have been completed without the constant encouragement and blind faith of my husband, David Kilcullen, my biggest fan and very best friend.

Acronyms

AAR	After-Action Review
AQI	Al Qaeda in Iraq
BCT	Brigade Combat Team
CAAT	Combined Arms Assessment Team
CALL	Center for Army Lessons Learned
CATP	Civil Affairs Training Program
CENTCOM	Central Command
CGSC	Command and General Staff College
CINC	Commander in Chief
CMTC	Combat Maneuver Training Center
COB	Civilians on the Battlefield
COIN	Counterinsurgency
CoP	Communities of Practice
CORDS	Civilian Operations Revolutionary Development Support
CORM	Commission on Roles and Missions
CPA	Coalition Provisional Authority
CSC	Command and Staff College
CT	Counterterrorism
CTC	Combat Training Center
DoD	Department of Defense
DPS	Defense Planning Scenarios
EUCOM	European Command
FID	Foreign Internal Defense
FM	Field Manual
FOB	Forward Operating Base
FSO	Foreign Service Officers
GAO	General Accounting Office
GPS	Global Positioning System
GW	Guerrilla Warfare
IDAD	Internal Defense and Development
IFOR	Implementation Force
IO	Information Operations
ISR	Intelligence, Surveillance, and Reconnaissance
IW	Irregular Warfare

JAG	Judge Advocate General
JFCOM	Joint Forces Command
JOC	Joint Operating Concept
JP	Joint Publication
JRTC	Joint Readiness Training Center
LIC	Low-Intensity Conflict
METL	Mission Essential Task List
MILES	Multiple Integrated Laser Engagement System
MNC-I	Multi-National Corps–Iraq
MNF-I	Multi-National Force–Iraq
MOOTW	Military Operations Other than War
MRE	Mission Rehearsal Exercise; Meal Ready to Eat
NCO	Noncommissioned Officers
NGO	Nongovernmental Organizations
NMS	National Military Strategy
NSS	National Security Strategy
NTC	National Training Center
OC	Observer Controller
OIF	Operation Iraqi Freedom
OPFOR	Opposing Force
OOTW	*See* MOOTW
PDD	Presidential Decision Directive
PE	Peace Enforcement
PK	Peacekeeping
PKI	Peacekeeping Institute
PKSOI	Peacekeeping and Stability Operations Institute
PME	Professional Military Education
PPBE	Planning, Programming, Budgeting, and Execution
PRD	Presidential Review Directive
PRT	Provincial Reconstruction Teams
QDR	Quadrennial Defense Review
RFI	Request for Information
S & R	Stability and Reconstruction
SAMS	School of Advanced Military Studies (Army)
SAW	School of Advanced Warfare (Marine Corps)
SME	Subject Matter Expert
SOCOM	Special Operations Command
SOF	Special Operations Forces
SOUTHCOM	Southern Command
TRADOC	Training and Doctrine Command
TTP	Tactics, Techniques, and Procedures
2-MTW	Two Major Theater War
USAID	United States Agency for International Development
USAIS	U.S. Army Infantry School
UW	Unconventional Warfare

Introduction: On the Front Lines with America's Nation Builders

On March 23, 2003, a battle began near the southern town of Nasiriya, Iraq. At least nine U.S. marines were killed, and many others were wounded. According to reports of the incident, marines had not seen such intense combat since the Vietnam War.[1] Unlike in the First Gulf War, when American and Iraqi armor clashed on the open desert, Iraqi combatants masqueraded as civilians and used "irregular" urban tactics to engage U.S. forces. Although frustrated by the initial confusion, the marines knew how to respond. One young sergeant claimed, "We're trained to do this."[2] By the end of the week, Nasiriya was under U.S. control, and Colonel Ronald Bailey, commander of Regimental Combat Team 2 of the Second Marine Division, was in charge. "All the sudden I was the mayor of eight cities," he explained. "I had no idea I would be responsible for getting the water running, turning on the electricity, and running an economy."[3] When asked what sort of formal military doctrine, guidance, or training he referenced to take on these tasks, he said he had read the 1940 version of the Marine Corps' *Small Wars Manual* in one of his officer courses but otherwise was trained for "real" war.[4]

Colonel Bailey's experience is far from unique. Today, U.S. military

---

1. Tim Pritchard, *Ambush Alley: The Most Extraordinary Battle of the Iraq War* (New York: Presidio Press, 2007).

2. Art Harris, *Growing up on the Front Line*, CNN.com, March 28, 2003 (accessed October 20, 2003).

3. Colonel Ronald Bailey, USMC, Commander, 2nd Regiment, 2nd Marine Division, personal interview by author, September 23, 2003, Arlington, VA.

4. USMC, *Small Wars Manual of the United States Marine Corps* (Washington, DC: U.S. Government Printing Office, 1940). To date, despite a number of recent attempts, this manual has not been updated since 1940. See chapter 4 for a discussion of this manual.

officers throughout Iraq and Afghanistan spend their days worrying nearly as much about sewer lines, water purification, trash collection, "rule of law," and jobs programs as they do about war fighting.[5] Despite common assertions that such tasks are not the job of the military, this book shows that fighting insurgents and common criminals while running elections, building roads, and kick-starting and governing cities have been perennial duties for U.S. soldiers and marines for over 230 years. They have always been frustrating, controversial missions that soldiers and marines were eager to forget. Thus, hard-learned lessons from so-called stability operations and counterinsurgency have rarely been captured into doctrine, training, or education, leaving each generation to rediscover what had been learned on the job before.

As America's military swings from Iraq back to Afghanistan in 2009, it is led by a generation of officers that have cut their teeth on the shores and in the cities of Panama, Somalia, Haiti, and the Balkans. In the seventh year of the campaign in Afghanistan—and in the 20th year since the fall of the Berlin Wall—the question is, will this time be different? Will this generation document and institutionalize its own experience and drive true organizational change to better prepare the future force for counterinsurgency or stability operations; or will it revert back to its traditional focus on "big war" as it has so many times before?

A popular view is that the U.S. military is culturally averse to conducting such "nonwar" missions and that this general aversion stymies its institutional capacity to learn from experience.[6] As this book demonstrates, however, this cultural explanation does not adequately account for the mixed record of military adaptation in such missions throughout history, such as the marines in the Caribbean in the 1930s or the development of civil affairs for post–World War II reconstruction; and it does not explain the level of learning that has occurred as a result of a series of "irregular" missions since the end of the Cold War and especially in response to operations in Iraq and Afghanistan.

While national security experts, pundits, policymakers, and military officers continue to debate the future of warfare and whether the pen-

---

5. Major General Peter Chiarelli, USA, Commander, First Cavalry Division, and Major Patrick R. Michaelis, USA, "Winning the Peace: The Requirement for Full Spectrum Operations," *Military Review* (July–August 2005).

6. Deborah D. Avant, "The Institutional Sources of Military Doctrine: Hegemons in Peripheral Wars," *International Studies Quarterly* 37 (1993); Robert M. Cassidy, *Peacekeeping in the Abyss: British and American Peacekeeping Doctrine and Practice after the Cold War* (London: Praeger, 2004); John A. Nagl, *Counterinsurgency Lessons from Malaya and Vietnam: Learning to Eat Soup with a Knife* (Westport, CT: Praeger, 2002).

dulum has swung too far away from "real" war, the institution continues to learn and adapt.[7] Contrary to conventional wisdom that the U.S. military is not a "learning organization," this book demonstrates that owing to a number of institutional changes put in place following the Vietnam War, the military has learned to learn from its own experience better than ever before. Today's military is simply not the same military that failed to learn and adapt after the failures of Vietnam. The next question is whether this military learning, in absence of true capacity development in other U.S. government agencies—such as the State, Justice, and Commerce departments and the United States Agency for International Development (USAID)—will be sufficient to meet the complex challenges of the 21st century.

Given this history, this book focuses on the record of U.S. military involvement in and adaptation to "stability and reconstruction operations" (S & R), which are considered by many to be something fundamentally distinct from "real war." It describes how officers have coped with the apparent mismatch between what they have prepared for (big war) and what they have repeatedly been asked to do (S & R). It examines how their operational experience has or has not translated to organizational learning—as reflected in new doctrine, education, and training—and why. By comparing the historical record to the current era, the book unravels the relative influence of organizational culture, political pressures, institutional structures, organizational processes, and, of course, leadership on the capacity of the U.S. military to learn, while also shedding light on the substantive elements that make this particular type of duty so difficult for military troops.

Defining the Mission

In order to understand the challenges associated with military learning for these "nonmilitary" missions, we need to define them. As figure 1 demonstrates, there is a vast array of confusing terms associated with this set of missions. This book will focus on *stability and reconstruction operations* (a.k.a. "stability ops") by U.S. military forces. Current U.S. military doctrine defines stability operations as follows: "various military missions, tasks, and activities conducted outside the United States in coordination with other instruments of national power to maintain or reestablish a safe and secure

---

7. Andrew Bacevich, *The Limits of Power: The End of American Exceptionalism* (New York: Metropolitan Books, 2008).

Fig. 1. Terms used to describe military operations. (Courtesy of Dr. William Flavin, Army Peacekeeping and Stability Operations Institute.)

environment, provide essential governmental services, emergency infrastructure reconstruction, and humanitarian relief."[8]

We might think of stability operations at the political and strategic level as civil-military interventions in the affairs of another nation-state for the purpose of reconstituting, assisting, or transforming the governing structures of that state. Thus, stability operations are a more competitive or violent version of the politically charged term *nation building*, as these complex interventions usually occur in societies under stress, with weak or failing governments. They are almost always dangerous missions, in which military forces (and their civilian partners) are challenged by an array of violent "spoilers," ranging from common criminals to warlords, drug lords, gangs, or insurgents. At the most competitive (and usually the most violent) end of this stability operations spectrum, *counterinsurgency* might be thought of as "an armed competition for governance."

---

8. Joint Publication 3-0, *Joint Operations*, February 13, 2008.

Not everyone will agree with the above description of the mission. For readers not steeped in the ever-changing military lexicon, simply note that there has historically been little consensus on the use of terms across the U.S. government and within the military when it comes to operations and missions that do not fit neatly into the traditional state-centric, forces-against-forces military model oriented around big battles. This large array of terms and the intellectual debates surrounding them reflect the perennial difficulty national security professionals have had in trying to make sense of these complex conflicts.

By the end of the Bush administration, the term *irregular warfare* was being used by a number of military professionals as an umbrella term to include all types of missions, including counterterrorism, insurgency, counterinsurgency, and stability operations. Although labeling these operations as a type of "warfare" made it easier for military leaders to embrace them as a core military mission and place more emphasis on how to organize, train, and equip for them, not everyone was convinced. Many of these "irregular wars" occur in environments where the United States is not officially at war and that require coordination with civilian agencies. Diplomats and other civilians rightly recoiled at the notion that their diplomatic or development activities would be labeled a type of "warfare," while some in the military rightly pointed out that these environments still often require the application of violence against hardened enemies, even if they are not formal military ones. This disagreement over terminology frustrates interagency coordination as well as the development of doctrine, a key instrument for military learning.[9]

Preview

This book begins, in chapter 1, with a rather academic review of the traditional literature on military change drawn from organizational theory and bureaucratic politics. Organizational culture is highlighted for its popular explanatory value in much of this literature. As a complement to traditional theories of military change, institutional learning is also presented as a means to examine the deeper institutional processes involved in military change. Chapter 2 provides a historical overview of the U.S. Army and the U.S. Marine Corps and their involvement in S & R. The purpose of this chapter is not to detail every military operation the U.S.

---

9. For a discussion of various terms see Janine Davidson, "Principles of Modern American Counterinsurgency: Evolution and Debate," Brookings Institution Counterinsurgency and Pakistan Paper Series, March 2009.

Army and Marine Corps have experienced in the past two centuries but, rather, to provide a general overview of the extent to which each service has been involved in nontraditional military missions, to describe the common characteristics of such operations, and to detail how the two institutions have or have not captured lessons from these episodes to better prepare troops for future similar contests. Because the learning record is quite mixed, chapter 2 highlights the various ways in which military structure, culture, and politics can prevent, permit, or promote bottom-up or top-down institutional adaptation and learning. In addition to providing insight on military adaptation and learning, chapter 2 also sheds light on the development of the current organizational culture of the U.S. military vis-à-vis S & R and war and provides the foundation for understanding the formal and informal organizational learning processes in place today.

Chapter 3 provides a more in-depth analysis of the origins and characteristics of the current military learning system and the political forces acting on it, as well as an overview of the military's apparent increasing involvement in these missions since the Vietnam War. This chapter reveals how post–Cold War strategic and political ambiguity and bureaucratic politics alternately stymied and promoted the learning process for these nontraditional missions. Chapter 4 describes how institutional changes designed to improve major warfighting capabilities, such as the combat training centers and the formal Lessons Learned Program, along with the development of the Light Infantry and the resuscitation of the Special Forces (both initiated on the margins of the 1980s military buildup), together enabled the military to conduct and adapt to these missions in the 1990s and beyond. Chapter 5 presents the numerous changes in S & R–oriented doctrine that developed throughout the 1990s and in response to Iraq and Afghanistan and describes the ways in which real-life operations, combined with internal structures, processes, and politics, influenced the production and dissemination of lessons learned and doctrine. By examining these post-Vietnam and post–Cold War developments, these chapters reveal the influence of experienced midlevel officers, senior-level officers, and civilian officials on the promotion and prevention of institutional learning and the parallel function of the formal training system as an engine of change.

Contrary to conventional wisdom, the current U.S. military learning system functions remarkably well for the security and combat-oriented aspects of S & R. This system functioned throughout the 1990s to improve the military's capacity to conduct peace and stability operations, even as the strategic and political debates and ambiguity over their actual roles and missions continued. Where the system falls short, however, is in preparing troops for the complex reconstruction tasks required in today's "stability and reconstruction" missions. With respect to building wells, roads, and schools and kick-starting governments and economies, doctrine, education, and training, as well as current policy proposals, do not adequately prepare troops on the ground conducting S & R today. Recent doctrinal developments, such as FM (field manual) 3-07 *Stability Operations,* described in chapter 5, reflect recent attempts to tackle this problem as well.

Finally, chapter 6 turns to the conflict in Iraq and describes the many factors that enabled General Petraeus and Ambassador Crocker to turn the tide of violence during the so-called Surge. In addition to leadership, the Surge benefited from a military that had undergone significant change. Colonel Bailey's military that invaded Iraq and toppled Saddam's military in 2003 and then found itself struggling to kick-start the society in order to win the peace was not the same American military Petraeus led in the Surge four years later. Somehow, in contrast to predictions to the contrary, it had learned.

CHAPTER 1

Military Learning and Competing Theories of Change

To Carl von Clausewitz, the father of modern military thought, military learning and military change were a simple matter: "If, in warfare, a certain means turns out to be highly effective, it will be used again; it will be copied by others and become fashionable; and so, backed by experience, it passes into general use and is included in theory."[1] In other words, if something works, militaries will change their doctrine and their practice accordingly. Although Clausewitz provides little explanation as to how this learning on the battlefield becomes organizational practice or why some armies learn while others do not, he claims that armies have at least three opportunities to learn—historical examples (of self and others), personal battlefield experience, and the experience of other armies.

In the case of the challenges facing the military today, the U.S. military has had the opportunity to learn in all three of the ways suggested by Clausewitz. Starting with frontier duty in the early 19th century and continuing to Iraq and Afghanistan today, the U.S. military has built schools, run local governments, monitored elections, and provided general law and order for war-torn societies both at home and abroad throughout its history. As chapter 2 describes, long before the peace operations of the 1990s or the "Phase IV" and counterinsurgency operations in Iraq and Afghanistan today, U.S. soldiers and marines performed myriad S & R tasks in the American South, the Philippines, the Caribbean, Europe, Japan, and Vietnam.

In addition to this historical and recent experience, the U.S. military

---

1. Carl von Clausewitz, *On War*, ed. Michael Howard and Peter Paret (Princeton: Princeton University Press, 1984).

also has a tradition of studying other militaries around the globe, demonstrating that the U.S. military is adept at learning from the experience of others. For example, the *Small Wars Manual* written by the Marine Corps in the 1930s reflects the lessons of the British from their 19th-century colonial wars as well as the Marine Corps' own experience in the Caribbean in the first decades of the 20th century.[2] Today, both the Army and the Marine Corps consult the British and other allies in preparing for urban operations, counterinsurgency, and peacekeeping.[3] Given this tradition, combined with their own long history of performing military operations other than war (MOOTW), Clausewitz would predict that the U.S. Army and the U.S. Marine Corps would be quite adept at performing them by now. Moreover, their techniques would differ little from each other or from those of their allies whose doctrine they had studied.

In contrast to Clausewitz, modern theories of military change suggest that militaries will have a difficult time innovating at all. A primary debate among scholars of military change is over the catalyst for innovation. Do militaries change on their own or in response to perceived threats, new technologies, or changes in the global system; or is some external stimuli required to force the organization and its leaders to "see the light" and adapt? If militaries are resistant to change, what does it take to influence their behavior from the outside? Under what conditions might efforts to force the military to innovate succeed or fail? If, on the other hand, militaries do change on their own, what (or who) influences the choices they make? Finally, whether the catalyst is internal or external, what explains the failure of militaries to change when needed?

In this literature, many posit that for various reasons, militaries need external actors to force innovation or change. The critical point of agreement among these scholars is that if left alone, the military would be unlikely to change or would otherwise tend toward inappropriate doctrine.[4] Scholars of military innovation draw on three overlapping cat-

---

2. Keith B. Bickel, *Mars Learning: The Marine Corps Development of Small Wars Doctrine, 1915–1940* (Boulder: Westview Press, 2001); Colonel C. E. Callwell, *Small Wars: Their Principles and Practice*, 3rd ed. (London: E. P. Publishing, 1976) (the first edition of *Small Wars* was published in 1896).

3. Multiple interviews with 101st soldiers, 1st Marines, 2nd Marines, Marine Corps Warfighting Lab, Army Peacekeeping and Stability Operations Institute, and the Army Center for Lessons Learned personnel.

4. Barry R. Posen, *The Sources of Military Doctrine: France, Britain, and Germany between the World Wars* (Ithaca: Cornell University Press, 1984); Jack L. Snyder, *Ideology of the Offensive: Military Decision-Making and the Disasters of 1914* (Ithaca: Cornell University Press, 1984).

egories of theory to explain why militaries often fail to adapt: organizational theory, bureaucratic politics, and organizational culture.[5] In each of these schools, there exist factors that make them either averse to change in general or inclined toward offensively oriented doctrine in particular.[6] Accordingly, each leads to different conclusions about how barriers to innovation might be overcome.

Organization Theory

Organization theory sees military organizations as highly resistant to change.[7] For organizational theorists, militaries resist innovation as a result of structural systems, norms, and standard operating procedures that together focus behavior toward particular outcomes. Graham Allison describes organizational behavior in this school "less as deliberate choices and more as outputs of large organizations functioning according to standard patterns of behavior." Moreover, organizational culture emerges from these routines that reinforce norms, and "the result becomes a distinctive entity with its own identity and momentum."[8] In this model, even when various actors within a military organization desire a change in strategy or doctrine, such structural mechanisms would likely mitigate against it. Thus, in order for such change to occur, the actual structures and processes that produce strategy and doctrine must be changed.

For today's military, many would point to the Pentagon's "planning, programming, budgeting, and execution" (PPBE) cycle as a key example of this phenomenon. In this complex process, the four services ideally submit budgets based on the leadership's strategic priorities as outlined

---

5. Graham Allison, *Essence of Decision: Explaining the Cuban Missile Crisis*, 2nd ed. (New York: Longman, 1999); Morton Halperin, *Bureaucratic Politics and Foreign Policy* (Washington, DC: Brookings Institution, 1974).

6. Carl H. Builder, *The Masks of War: American Military Styles in Strategy and Analysis*, Rand Corporation Research Study (London: Johns Hopkins University Press, 1989); Elizabeth Kier, *Imagining War: French and British Military Doctrine between the Wars* (Princeton: Princeton University Press, 1997); Stephen Peter Rosen, *Winning the Next War: Innovation and the Modern Military* (Ithaca: Cornell University Press, 1991); Steven Van Evera, "The Cult of the Offensive and the Origins of the First World War," *International Security* 9, no. 1 (1984).

7. Allison, *Essence of Decision;* Deborah D. Avant, "The Institutional Sources of Military Doctrine: Hegemons in Peripheral Wars," *International Studies Quarterly* 37 (1993); James March and Johan Olsen, "The New Institutionalism: Organizational Factors in Political Life," *American Journal of Political Science* 78, no. 3 (1984); Halperin, *Bureaucratic Politics and Foreign Policy;* Snyder, *Ideology of the Offensive*.

8. Allison, *Essence of Decision*.

in the National Defense Strategy or the Quadrennial Defense Review, which should in turn reflect the immediate and projected needs of the warfighters, as articulated somehow by the combatant commanders around the world. In reality, budgets often seem out of touch with both top-down priorities and bottom-up requests, reflecting a mysterious disconnect between bottom-up learning, strategic direction, and budgeting.

Pushing change from above requires the strategic manipulation of the system at key "nodes" in this PPBE process. For example, the choice of defense planning scenarios (DPSs) or war games and analytical scenarios used for capability analysis inside the Pentagon during the budget cycle or during the development of the Quadrennial Defense Review can have cascading effects on what the services think they are required to program and budget for. A war game that presumes major conventional warfare against a fictitious peer competitor, for instance, would lead the Army to buy big tanks and the Air Force to buy high-tech fighter planes capable of air-to-air combat. If, on the other hand, the directed scenarios emphasize military operations such as counterinsurgency or humanitarian intervention, the justification for these major weapons systems gives way to other capabilities: for example, special operations; language skills; intelligence, surveillance, and reconnaissance (ISR) equipment; and humanitarian relief capabilities.

For analysts trying to measure change, however, the budget can often be misleading. The same size budget for military education, training, or doctrine development, for instance, may look the same on the outside, but the character and content of these programs may have adapted dramatically. The way in which experience affects how the military thinks, trains, and learns, as evidenced in its doctrine, training, and education (more so than its budgeting and weapons procurement), is the primary focus of this book.

Despite such barriers to innovation, organization theory makes room for military change in response to three catalysts: (1) external pressure, (2) the opportunity or need to grow and/or survive, and (3) failure.[9] In the first instance, external civilian leadership—Congress, the president, or even the secretary of defense—would be the source of change. In the

---

9. Ibid.; Anthony Downs, *Inside Bureaucracy* (Boston: Little, Brown, 1967); Posen, *The Sources of Military Doctrine*; Chris Argyris and Donald Schon, *Organizational Learning II: Theory, Method, and Practice* (Reading, MA: Addison-Wesley, 1996); Richard M. Cyert and James G. March, *A Behavioral Theory of the Firm*, 2nd ed. (Malden, MA: Blackwell, 1992); Douglass C. North, *Institutions, Institutional Change, and Economic Performance* (New York: Cambridge University Press, 1990).

second case, a military might change in order to acquire more resources or influence.[10] The last of the organizational theory categories, failure, is more intuitive with respect to the military. In this case, as militaries face new technologies or tactics in use by an enemy on the battlefield, they are forced to adapt. In a very Clausewitzian sense, militaries that do not adapt do not survive. What is less clear in these cases is *how* the change takes place. What are the mechanisms and institutional processes that enable this change?

Bureaucratic Politics

Like organization theory, bureaucratic politics theory might predict similar outcomes when the military is viewed as the amalgam of myriad subgroups and branches as well as one agency among others within the U.S. government.[11] In this model, most commonly summarized by the adage "Where you stand depends on where you sit," military leaders, like leaders of other large organizations, seek to promote the importance of their organization and to preserve the organization's distinct organizational "essence." Morton Halperin defines this essence as "the view held by the dominant group in the organization of what the missions and capabilities should be."[12] In this model, roles and missions that challenge this essence will be rejected, unless such roles are seen to enhance the importance and influence of the organization. Thus, in the bureaucratic politics model, we would expect the military to resist stability operations missions unless they can be viewed as somehow supporting the organization's essence or somehow increasing its stature or relevance.

Because, as we will see, the predominant view of the U.S. military as a whole is that its role is to "fight and win the nation's wars," it would seem unlikely that the military would embrace counterinsurgency (COIN) or stability operations as its role. In the post–Cold War environment, however, when it seemed that the implosion of the Soviet enemy meant massive downsizing and a diminished role for the U.S. military, the bureau-

---

10. Indeed, the perception that the United States might no longer need large numbers of forces in Europe to counter the Soviets did shrink military budgets in the early 1990s. Many in the military at that time perceived such calls for military downsizing as an existential threat.

11. Allison, *Essence of Decision;* Michael Altfeld and Gary Miller, "Sources of Bureaucratic Influence: Expertise and Agenda Control," *Journal of Conflict Resolution* 28, no. 4 (1984); Halperin, *Bureaucratic Politics and Foreign Policy.*

12. Halperin, *Bureaucratic Politics and Foreign Policy;* Jerel Rosati, "Developing a Systematic Decision-Making Framework: Bureaucratic Politics in Perspective," *World Politics* 33 (January 1981).

cratic politics model, like organization theory described above, would lead to the competing hypothesis that the military would embrace what it then called "MOOTW" as its new raison d'être, as a means to maintain its organizational influence. Likewise, inside the military establishment, the bureaucratic politics model would lead to an additional hypothesis that subgroups such as the Special Forces or the Light Infantry, whose capabilities are uniquely suited to the challenges of MOOTW missions, might also advocate for increased emphasis on this role. Indeed, this is the behavior that began to emerge in the 1990s and especially once operations in Afghanistan and Iraq began.

Organizational Culture and Military Change

Scholars who focus on organizational culture often use different terms with slightly different emphases to describe a similar phenomenon. For example, Elizabeth Kier defines organizational culture as follows: "the set of basic assumptions and values that shape shared understandings, and the forms or practices whereby these meanings are expressed, affirmed, and communicated to the members of an organization."[13] Closely related to organizational culture is Morton Halperin's concept of organizational essence, described above. Other scholars focus on institutional memory, "the conventional wisdom of an organization about how to perform its tasks and missions."[14] Richard Downie clarifies this concept further by stating, "In a sense, institutional memory is what older members of an organization know and what new members learn through a process of socialization."[15] Finally, Carl Builder presents the theory of organizational personality: "a 'face' that can be remembered, recalled, and applied in evaluating future behavior [of a military service]."[16] These terms, which are often used interchangeably, are for some scholars the key to understanding most differences in military behavior.

The centrality of culture and the relationship of these concepts for military organizations are articulated clearly by Lieutenant General

---

13. Elizabeth Kier, "Culture and Military Doctrine: France between the Wars," *International Security* 19, no. 4 (1993). See also James Q. Wilson, *Bureaucracy* (New York: Basic Books, 1989): "Every organization has a culture, that is, a persistent, patterned way of thinking about the central tasks of and human relationships within an organization."

14. John A. Nagl, *Counterinsurgency Lessons from Malaya and Vietnam: Learning to Eat Soup with a Knife* (Westport, CT: Praeger, 2002).

15. Richard Duncan Downie, *Learning from Conflict: The U.S. Military in Vietnam, El Salvador, and the Drug War* (Westport, CT: Praeger, 1998).

16. Builder, *The Masks of War.*

Theodore Stroup: "The Army's culture is its personality. It reflects the Army's values, philosophy, norms, and unwritten rules. Our culture has a powerful effect because our common underlying assumptions guide behavior and the way the Army processes information as an organization."[17] General Stroup goes on to claim, "Our Army culture, however, can also be a liability when it is inappropriate and does not contribute to the Army's overall goals."[18] But where does culture come from, and what, if anything, can be done to overcome its powerful and potentially negative, reactionary influence?

To understand the origins of organizational culture, most scholars look to an organization's history. As Carl Builder explains, recent and historical experience is key to understanding the origins of organizational personality.

> Like all individuals and durable groups, the military services have acquired personalities of their own that are shaped by their experiences and that, in turn shape their behavior. And like individuals, the service personalities are likely to be significantly marked by the circumstances attending their early formation and their most recent traumas.[19]

That early experiences have a disproportionately formative influence on the personality and behavior of an institution (or a person) resonates in learning theory as well.[20] Moreover, as John Nagl observes, "organizational culture also plays a critical role in determining how effectively organizations can learn from their own experiences."[21] Thus, an organization's history affects the development of the organization's personality, which in turn affects the ability of the organization to learn from new experience. This iterative relationship between experience, culture, and learning suggests that culture can be an incredibly determinate factor in the behavior of an organization. Accordingly, for a number of students of military performance in MOOTW, organizational culture is the key to understanding success or failure in new operating environments.

---

17. Theodore G. Stroup, Jr., "Leadership and Organizational Culture: Actions Speak Louder than Words," *Military Review* 171, no. 1 (1996).
18. Ibid.
19. Builder, *The Masks of War.*
20. Robert Jervis, *Perception and Misperception in International Politics* (Princeton: Princeton University Press, 1976); Daniel H. Kim, "The Link between Individual and Organizational Learning," *Sloan Management Review* 35, no. 1 (1993); D. A. Kolb, *Experiential Learning* (Englewood Cliffs, NJ: Prentice-Hall, 1984).
21. Nagl, *Counterinsurgency Lessons from Malaya and Vietnam,* 11.

Integrated Theories of Military Change

Most scholars take an integrated approach to explaining military change. Modern literature on military innovation focuses on either external or internal sources of military innovation and borrows from the theories and perspectives outlined above. In combining many of these approaches, scholars also frequently highlight the importance of organizational culture and civil-military relations in promoting or preventing innovation. A common theme in this literature is how uncommon or difficult it is for militaries to change. Defining the conditions under which barriers to change may or may not be overcome is the goal of such research.

The first category of military change literature consists of scholars who posit that militaries need external actors to force innovation.[22] Explanations for this failure to adapt include organizational or bureaucratic barriers,[23] cultural factors,[24] a predilection for offensively oriented doctrine,[25] or some combination of these elements. In this literature, the civil-military dynamic is critical, as civilian leaders must interact with their military counterparts to drive the organization to innovate.[26]

The scholar most commonly attributed to this "civilian-intervention" approach is Barry Posen. In *The Sources of Military Doctrine,* Posen demonstrates that militaries resist change or otherwise cling to offensive doctrines in accordance with organization theory. Although military leaders may consider adjusting their doctrines "when threats become sufficiently grave," it is mostly civilians who, in accordance to balance of power theory, identify the need for new military doctrine and intervene to force change on the military. Their success in pushing change from outside the military depends on the delicate nature of civil-military relations.[27] In the United States, efforts by civilians to reorient or transform the military (i.e., President Kennedy's push for counterinsurgency doctrine during the Vietnam conflict or President Clinton's push for interagency coordination for peace operations) have often been uphill battles. Theories developed by scholars such as Deborah Avant and Stephen Rosen

---

22. Snyder, *Ideology of the Offensive.*
23. Allison, *Essence of Decision;* Halperin, *Bureaucratic Politics and Foreign Policy.*
24. Builder, *The Masks of War;* Kier, *Imagining War;* Rosen, *Winning the Next War.*
25. Rosen, *Winning the Next War;* Snyder, *Ideology of the Offensive;* Van Evera, "The Cult of the Offensive and the Origins of the First World War."
26. Posen, *The Sources of Military Doctrine;* Paul R. Viotti, "Introduction: Military Doctrine," in *Comparative Defense Policy,* ed. Frank B. Horton III, Anthony C. Rogerson, and Edward L. Warner III (Baltimore: Johns Hopkins University Press, 1974).
27. Posen, *The Sources of Military Doctrine.*

provide different explanations for why militaries are not always responsive to such civilian efforts.[28]

Stephen Rosen suggests that career structures within military organizations reward officers who follow traditional paths. Armor officers, for example, get promoted for mastering armor doctrine and being good tank drivers, not for promoting new warfighting tactics that emphasize the benefits of light forces and smaller vehicles in urban terrain. A recent illustrative case is that of the officers who wrote the Army's manuals for peace operations in the 1990s. They claimed they spent time on the project even though they knew that it would "kill" their chances of getting promoted.[29] This supports Rosen's assertion that "maverick" officers who advocate change from within a conservative military organization often suffer professionally for their efforts.

Rosen focuses on military culture as the source of this resistance and claims that military leadership can drive change by changing the career incentives to reward young officers who operate outside the traditional systems.[30] Such an organizational climate would reward initiative and creativity but would also need to be more tolerant of mistakes.[31] In this case, change flows with the new generation and therefore occurs slowly. Deborah Avant agrees that career incentives are an important element, but she adds another layer to the analysis: the political system.

Avant, who examines both the internal elements of military organizations and the political systems in which they reside, challenges the concept that militaries are inherently resistant to change.[32] She notes that the British military adapted to the exigencies of the Boer War without civilian intervention, while the American Army failed to adapt during the Vietnam War—despite significant urging by civilian leadership to change course. She claims that the American system of split civilian control over the military (congressional and executive) enabled the U.S. Army to resist intervention, while its culture reinforced a distaste for

---

28. Rosen, *Winning the Next War*.

29. Colonel George Oliver, U.S. Army, former Director, Army Peacekeeping Institute, numerous personal interviews by author, October 2003 to June 2005, Washington, DC.

30. Civilians can also intervene in the career structure to drive change. A modern illustration of this is the 1986 Goldwater-Nichols legislation. By tying career progression to having served in a "joint" assignment (i.e., an Army officer spending a few years with the Navy) the U.S. Congress was able to promote more integration, interoperability, and cultural understanding among the three services.

31. A common criticism in military circles (and other organizations) is that a "zero-defect" mentality creates a risk averse culture where creativity is stymied.

32. Deborah Avant, *Political Institutions and Military Change: Lessons from Peripheral Wars* (Ithaca: Cornell University Press, 1994); Avant, "The Institutional Sources of Military Doctrine."

counterinsurgency. In contrast, the British system provided more uniform military oversight, which had enabled the military to develop in a culture that was more flexible and less reactive. Thus, Avant proposes that different government structures create different patterns of civilian oversight that, in turn and over time, influence military culture and give militaries varying degrees of flexibility or bias toward change.

Others join Rosen and Avant in focusing on internal organizational, cultural, or structural elements as the sources of or impediments to change. Kimberly Zisk suggests that militaries are concerned with survival and will respond on their own to "threats"—both on the battlefield and on the domestic political front. She credits militaries for reacting on their own to changes in the environment of threat to national security (as defined by changes in the enemy military's doctrine), but she also claims that militaries are even more concerned with domestic political challenges at home. Reflecting theories of bureaucratic politics, she claims that militaries will respond more vigorously to threats to their budgets or resources than to external cues to change their warfighting doctrines.[33]

With respect to culture, Elizabeth Kier takes a slightly different approach. Whereas Rosen focuses on the culturally driven obedience to a conservative career structure as the primary impediment to military change, Kier notes that militaries are also bound by constraints set by civilian leaders—constraints that are internalized into the military culture itself. Like Avant, her approach attempts to bridge the gap between structural theories for military doctrine and cultural ones. Civilian leaders create the structures in which militaries operate, and military leaders learn to operate within those structural constraints. Her emphasis on organizational survival and military culture means that her study challenges those who locate the source of military resistance on an inherent military preference for offensive doctrines: "It may not be the offensive aspect of their doctrine that the military seeks to safeguard, but instead some part of its traditional way of doing things whose preservation is, for these officers, integral to the successful execution of their mission."[34]

In sum, the literature on military change outlined above sheds light

---

33. Kimberly Marten Zisk, *Engaging the Enemy: Organization Theory and Soviet Military Innovation 1955–1991* (Princeton: Princeton University Press, 1993).

34. Kier's work responds to a popular debate among scholars over the source of offensive versus defensive doctrines (strategies). Although such work was focused on explaining the character of and derivation of doctrine, the literature contributed to theories of military change and civil-military relations in general. Van Evera, "The Cult of the Offensive and the Origins of the First World War"; Posen, *The Sources of Military Doctrine;* Snyder, *Ideology of the Offensive.*

on the difficult process of military innovation. Most agree that change does not occur easily or automatically. Militaries often tie their cultural identities to specific roles or have career structures that fail to reward (or even punish) new ways of thinking. Militaries that do change from within are described as having more "flexible" cultures that somehow promote innovative thinking and then manage to translate these new ideas into doctrine. Although this literature provides a useful framework for initial inquiry, questions remain unanswered. How exactly do new ideas take hold? How are they transferred to the organization as a whole? In short, what are the processes by which militaries learn and change? What, if anything, can be done to enhance the ability of a military organization to learn from its experience and change its doctrine and practices? For this, we turn to theories of organizational learning.

Military Change as Organizational Learning

The growing body of organizational learning theory provides a framework for understanding how militaries learn and adapt.[35] For this research, Richard Downie's definition provides a reference point for understanding organizational learning: it is "a process by which an organization (such as the U.S. Army) uses new knowledge or understanding gained from its experience or study to adjust institutional norms, doctrine, and procedures in ways designed to minimize previous gaps in performance and maximize future successes."[36] A normal process for organizational learning requires that learning begin on the individual level. Individuals within the organization first recognize, either through experience or personal study, the need for change. They then act within the norms and procedures of that organization to stimulate organizational change. This process changes the "institutional memory" of the organization. The new institutional memory is vulnerable to the same process given new learning by new individuals. Institutional learning theorists have different models to describe this loop, varying in complexity, critical factors, and numbers of steps.

According to organizational learning theory, some organizations may

---

35. Jervis, *Perception and Misperception in International Politics;* Kim, "The Link between Individual and Organizational Learning"; James H. Lebovic, "How Organizations Learn: U.S. Government Estimates of Foreign Military Spending," *American Journal of Political Science* 39, no. 4 (1995); James G. March and Johan P. Olsen, "Organizational Learning and the Ambiguity of the Past," in *Ambiguity and Choice in Organizations,* ed. James G. March and Johan P. Olsen (Bergen: Universitets forlaget, 1979); Peter M. Senge, *The Fifth Discipline: The Art and Practice of the Learning Organization* (New York: Currency Doubleday, 1990).

36. Downie, *Learning from Conflict.*

be more culturally predisposed to collect information and learn from experience than others. While some organizations actively promote the collection and dissemination of new information, others rigidly adhere to standard operating procedures and ignore new information—especially if that information challenges existing paradigms and norms. For example, John Nagl claims that in contrast to the rigid system of the U.S. Army in Vietnam, the "British Army was in fact a 'learning institution' during the [Malayan campaign] as a result of its organizational culture."[37] His study places a premium on the influence of organizational culture as the key element in an organization's capacity to learn.

Although organizational learning theorists also focus on the role of culture, much of the work in this field is focused on creating institutions, initiating processes, and making structural changes to an organization in order to actively promote learning. In fact, some organizational learning theorists, such as Peter Senge, author of the best-selling book *The Fifth Discipline*, advise organizations on how they can become "learning institutions."[38] This would suggest that there are concrete actions that can be taken by an organization's leadership to overcome cultural resistance to organizational learning. It suggests that structures and processes matter as much as, if not more than, culture.

This research tests this assumption by paying particular attention to how organizational changes in structure and processes over time, such as the creation of formal military schools and the introduction of war-planning processes, influenced the learning systems of the U.S. military. For example, how did the fledgling structure of the U.S. Army on the frontier compare to that of the U.S. Marine Corps in the Banana Wars and the U.S. Army in the interwar years in its formal ability to gather, create, and disseminate information? Likewise, how much does the sheer size of an institution affect its ability to disseminate information, learn, and adapt?

In the case of the U.S. Marine Corps, General James Mattis, former commander of the First Marine Division in Iraq, claimed, "I learned more about life in this profession at happy hours and reading the Gazette than I did in all my training and PME [professional military education]."[39] This statement reflects the cultural disposition (and the widely accepted reputation) of his organization to informally share knowledge. Indeed, this is a common theme in the Marine Corps, which

---

37. Nagl, *Counterinsurgency Lessons from Malaya and Vietnam*, 6.
38. Senge, *The Fifth Discipline*.
39. Lieutenant General James Mattis, USMC, personal interview by author, June 15, 2005.

has traditionally been more reluctant to record and rely on formal doctrine compared to the Army. It is possible that the Marine Corps, being significantly smaller than the Army, has historically had a simpler time transferring institutional memory by such informal means and has thus not had to develop formal systems to promote organizational learning. The question is, are the systems and traditions that the Corps developed over time still reliable now that the institution has grown? The U.S. Marine Corps is the smallest U.S. military service, yet today, at 175,000 members (and scheduled to grow more in the 2010 budget), it is still significantly larger than the British Army of 105,000.

For the U.S. Army, as we will see, a number of formal procedures for war planning that were adopted in the interwar years greatly affected the organization's learning capacity. By mandating historical review in order to generate war plans, these procedures actively promoted a reflective learning process resulting in increased attention being paid to the problems of stability and reconstruction (what was then referred to as "military governance"). Even more profound were the organizational changes enacted following the Vietnam War. Leaders in the post-Vietnam generation actively set out to change the learning culture of the Army in order to overcome the pathologies of the Vietnam War.[40] These structural and procedural changes, such as the new high-tech combat training centers, the process of after-action review, and the formal Center for Army Lessons Learned, consciously applied concepts from organizational learning theory in a deliberate attempt to gather information "from the field."[41] According to former Army chief of staff General Gordon Sullivan, the Army leadership consciously sought to create a "learning organization," even seeking the advice of learning theorists such as Peter Senge.[42]

---

40. The failures of the U.S. military to adapt and the ongoing debates over the Vietnam War are outlined in numerous studies. Avant, *Political Institutions and Military Change;* Robert M. Cassidy, "Why Great Powers Fight Small Wars Badly," *Military Review* (2000); Eliot Cohen, "Constraints on America's Conduct of Small Wars," in *Conventional Forces and American Defense Policy,* ed. Steven E. Miller (Princeton: Princeton University Press, 1986); Conrad Crane, "Avoiding Vietnam: The U.S. Army's Response to Defeat in Southeast Asia," *Strategic Studies Institute* (2002); Downie, *Learning from Conflict;* Andrew F. Krepinevich, Jr., *The Army in Vietnam* (Baltimore: Johns Hopkins University Press, 1986); Nagl, *Counterinsurgency Lessons from Malaya and Vietnam.*

41. James Kitfield, *Prodigal Soldiers: How the Generation of Officers Born of Vietnam Revolutionized the American Style of War* (New York: Brassey's, 1997); Gordon Sullivan and Michael V. Harper, *Hope Is Not a Method: What Business Leaders Can Learn from America's Army* (New York: Random House, 1996).

42. General Gordon Sullivan, USA, personal interview by author, February 25, 2005, Arlington, VA.

The question is, how did these structural changes interact with the other very real cultural, organizational, and bureaucratic political forces highlighted by military innovation theorists? How did they operate with respect to MOOTW? The remainder of this chapter provides a foundation for examining these questions by outlining the basic concepts and terms used in organizational learning theory.

Organizational Learning Theory Defined

Organizational learning theory builds on learning theory applied at the individual level but seeks to describe the processes by which learning occurs within and throughout an organization. What distinguishes organizational learning from learning by individuals—or even groups of individuals within an organization—is that learning acquired through this process remains even when personnel change. Thus, to determine if organizational learning has occurred, we must analyze not only the processes in place during an event for contemporaneous learning or adaptation but the evidence of learning remaining after the event. For the latter, we look for evidence that tactics, techniques, and procedures learned in action at one point in time are applied at the start of action at a later date. Organizational learning concepts—such as the learning cycle; informal, experiential, and generational learning; informal networks; and communities of practice—provide a framework for examining the learning systems of military organizations.

The Learning Cycle

In the organizational learning cycle, organizational learning takes place when knowledge is acquired by individuals—at any level—and then disseminated to the organization as a whole. There are many different ways theorists have to describe this loop. As figure 2 shows, in its most basic form, the cycle contains three major points: scan—interpret—act.[43] A "learning organization" will have formal processes to promote each of these steps. *Scanning* involves the focused effort to capture lessons from action. Ideally, an organization would have "collection" processes in place that target opportunities to collect data and a method for sorting

---

43. Lloyd Baird, John C. Henderson, and Stephanie Watts, "Learning from Action: An Analysis of the Center for Army Lessons Learned (CALL)," *Human Resource Management* 36, no. 4 (Winter 1997); R. L. Daft and K. E. Weick, "Toward a Model of Organizations as Interpretation Systems," *Academy of Management Review* 9, no. 2 (1984).

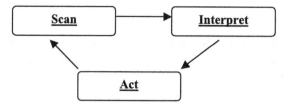

Fig. 2. Simple learning cycle. (Data from Lloyd Baird, John C. Henderson, and Stephanie Watts, "Learning from Action: An Analysis of the Center for Army Lessons Learned (CALL)," *Human Resource Management* 36, no. 4 [Winter 1997]: 385–95.)

and managing the information. To *interpret* is to make sense of the information, to track themes and trends over time, to identify cause and effect, and to synthesize and codify the information in a format that can be disseminated. Finally, in the *act* phase, the new knowledge is actively disseminated, and the lessons are applied in action at the new opportunity. In a learning organization, "all three phases happen in an ever-repeating cycle . . . Action leads to further scanning, interpreting, and acting."[44]

A critical element of the learning process is the elimination of ambiguity in order for the cycle to continue.[45] There must be clarity and consensus regarding the events that occurred (what happened or what is happening), what those events mean (why this matters to our organization), and what the proper course of action should be (what should be done about it).[46] Because individuals, for a variety of reasons, interpret events in different ways, the potential for ambiguity and disagreement throughout this cycle is acute.

Although organizational learning cycles have no true beginning or end, the organizational learning process depends on learning that occurs at the individual level. According to Peter Senge, "organizations learn only through individuals who learn. Individual learning does not guarantee organizational learning. But without it no organizational learning occurs."[47] Thus, in order to understand how individuals are able to reach consensus within groups and how individual and group learning influences organizational learning, it is important to review the cognitive learning processes at the individual level and group levels.

---

44. Baird, Henderson, and Watts, "Learning from Action."
45. March, "Organizational Learning and the Ambiguity of the Past."
46. Downie, *Learning from Conflict;* March, "Organizational Learning and the Ambiguity of the Past"; Sullivan and Harper, *Hope Is Not a Method.*
47. Senge, *The Fifth Discipline,* 139.

Experiential Learning

Individuals can learn either formally (through institutionally provided education and training) or informally (through self-study). In each of these cases, learning can occur experientially (through "hands-on" activities) as well as through intellectual reflection (reading, listening, thinking). Informal learning that occurs through experience is especially important for military change—particularly unplanned events and surprises. As Clausewitz suggested, when troops encounter phenomena in the field for which they have not been trained, they have an opportunity to learn through personal experience. How well individuals learn at that point is contingent on a number of factors, such as their previous knowledge, training, and education—all of which may help or hinder their ability to make sense of an unfamiliar situation and adapt. Additionally, we can expect different people to interpret events differently—and often incorrectly—depending on their previously held beliefs, assumptions, and worldviews.[48]

Generational Learning

The role of experience and worldviews resonates in theories of "generational change." In this model, experiential learning at a younger age is considered more critical to the formulation of individual worldviews. This learning is facilitated by the sharing of knowledge among the members of one's generation and has a delayed impact on the behavior of the organization. As Jack Levy explains, "the models' key hypothesis is that the shared experiences (and interpretations of them) of people at certain critical stages of their personal, intellectual, or political development have a powerful and enduring impact on their beliefs about the world, so that different generations learn different lessons."[49] Once a new generation whose early professional experience differed greatly from the one before it gains decision-making authority in an organization, we might expect new policies to be applied and organizational change to occur. For example, the many organizational and doctrinal

---

48. Graham Allison, *Essence of Decision: Explaining the Cuban Missile Crisis* (Boston: Little, Brown, 1971); Jervis, *Perception and Misperception in International Politics;* Kolb, *Experiential Learning;* Barbara Leavitt and James G. March, "Organizational Learning," *Annual Review of Sociology* 14 (1988). Likewise, political psychology research into foreign policy decision making suggests that worldview "lenses" or closely held beliefs influence leaders' decisions more than empirical observation of events.

49. Jack S. Levy, "Learning and Foreign Policy: Sweeping a Conceptual Minefield," *International Organization* 48, no. 2 (1994): 290.

changes that occurred following the Vietnam War were, for many analysts and military leaders, a direct result of a new generation, with shared experiences, taking control.[50]

In generational learning, the critical link is how individuals at similar points in their personal development *collectively* make sense of their experience and whether or not they are subsequently able to apply their lessons on an organizational level. As discussed below, "informal networks" and "communities of practice" (CoPs) can facilitate collective learning and help mitigate the ambiguity discussed above, but without the correct structure and processes in place, such learning will not be sufficient to ensure organizational change.

Informal Networks and Communities of Practice

Both "informal networks" and "communities of practice" are comprised of individuals who voluntarily participate in order to share information. In an informal network, individuals learn from each other's experiences by sharing problems, ideas, and solutions. For example, a group of runners might share diet and training tips to improve their workouts and individual race results. Communities of practice differ from informal networks in that CoP members are linked by a greater sense of culture, identity, and purpose. They are more than a group of individuals hoping to learn from each other to improve their individual performance. They are a community of experts committed to improving the practice within their profession.

Members of CoPs share new ideas and identify "best practices" for their professions. They communicate in similar ways as informal networks but might also meet at formal conferences and publish papers focused on improving the profession. Both informal networks and CoPs facilitate the spread of information within an organization, but CoPs can have a more targeted role in the organizational learning cycle. Organizational learning can be enhanced by informal networks and CoPs by helping individuals make shared sense of their experience.

In the organizational learning cycle, the transfer of individual learning to collective learning requires that a consensus be reached among members of the organization regarding the best course of action. The collective sharing of information through informal networks and CoPs facilitates this process but does not ensure that a consensus will be reached. If members of the group do not agree, organizational learning

---

50. Kitfield, *Prodigal Soldiers;* Sullivan and Harper, *Hope Is Not a Method.*

can be blocked. Such disagreement is even more likely to occur when changes are proposed that challenge the "governing variables" of the institution.

The Role of Leadership

The fact that organizational learning theory has become popular among corporate leaders reflects the critical role that leaders play in preventing, promoting, or permitting learning to occur in their organization. As we will see, with a good understanding of learning theory (or even simple intuition about what might work), savvy leaders can create structures and processes within their organizations that are expressly designed to facilitate organizational learning. Such leaders create systems that actively capture lessons from experience, allow the cross-fertilization of new ideas, and promote the dissemination of new knowledge. Changes made by Army leadership in the decades following the Vietnam War to the U.S. Army's structures and processes for collecting and disseminating new knowledge through experience are presented in this study as an example of such a leader-directed transformation.

On the other hand, leaders can also stymie organizational learning by intervening in existing learning processes or by creating processes that hinder bottom-up communication or fail to capture and disseminate new knowledge. Thus, leaders can fundamentally alter the learning culture of an organization through the good or bad design of processes and systems; and they can also stymie existing processes through targeted actions at critical points in the system that block learning processes.

In sum, organizational learning is a complex cycle involving a number of interconnected processes. Organizations that fail to learn are often stymied by factors such as cognitive beliefs by powerful leaders, organizational incentive structures that discourage creativity, or structural processes that block the transmission of knowledge. These factors reflect similar themes in organization theory and bureaucratic politics, on which much of the military innovation literature is based. While individual learning is necessary, it is not sufficient for organizational learning to occur. As one author claimed, organizations must possess "the right culture, the knowledge itself, and access to the knowledge" in order to learn.[51] The ways in which organizational culture, formal structures, and organizational processes influence military learning—and are iteratively influenced by that learning—are examined in the chapters that follow.

---

51. Sullivan and Harper, *Hope Is Not a Method*, 36.

CHAPTER 2

Two Centuries of Small Wars and Nation Building

This obviously wasn't peace—yet it didn't quite seem war. This was neither Kenya nor Indo-China. The system, as usual, told us nothing.[1]

—SECOND LIEUTENANT OLIVER CRAWFORD,
THE DOOR MARKED MALAYA

Writing of his experiences in Somalia in 1992, Pentagon reporter Tom Ricks observed, "This was the first U.S. brush with 'peacemaking'—a new form of post–Cold War, low intensity chaos that is neither war nor peace, but produces enough exhaustion, anxiety, boredom, and confusion to feel much like combat."[2] Indeed, many observers of the U.S. military's ill-fated mission to Somalia thought that a new type of low-level conflict had emerged—and they were not so sure that U.S. forces were an appropriate remedy. History indicates, however, that such missions are far from new for U.S. troops. Messy and confusing operations in Somalia and Haiti in the 1990s were not unlike frontier army duty in the 1800s, occupation in the American South after the Civil War, missions to the Philippines and Cuba after the Spanish-American War, or the so-called Banana Wars of the early 20th century. In fact, all of these experiences have far more in common with 21st-century operations in Iraq and Afghanistan than do the "major combat operations" for which the military has been organized, trained, and equipped throughout the majority of the 20th century. As this chapter reveals, the U.S. Army and

---

1. Oliver Crawford, *The Door Marked Malaya* (London: Rupert Hart-Davis, 1958); reprinted in 1989 as FMFRP 12-28 USMC.
2. Tom Ricks, *Making the Corps* (New York: Scribner, 1998), 17.

Marine Corps have a long history of conducting such missions and a mixed record of learning from their experiences.

The purpose of this chapter is not to detail every military operation the U.S. Army and Marine Corps have experienced in the past two centuries but, rather, to provide a general overview of the extent to which the military has been involved in nontraditional military missions, to highlight the common characteristics and controversies associated with such operations, and to detail how the two institutions have or have not captured lessons from these episodes to better prepare troops for future similar duties. The organizational and political narratives that evolved from this rich history have informed the modern military's mind-set and its understanding of its "proper" role. While the actual operational experience provided substantive lessons for current and future operations, the narratives that evolved, both within the military and throughout American society, presented cultural and political obstacles to organizational change.

This chapter presents a mixed record of success and failure over two centuries as the American military learned, relearned, and institutionally forgot again and again the types of complex stability operations and counterinsurgency tasks it finds itself conducting again today. The chapter outlines the informal mechanisms by which troops coped with unfamiliar missions and how they shared information and adapted contemporaneously when their institutions failed to prepare them adequately. More formal attempts to convert this adaptation into longer-term organizational learning, such as the Marine Corps' experience in crafting the *Small Wars Manual* in the 1930s and the War Department's development of systematic planning processes and schools for military government and civil affairs during the interwar years, demonstrate the way in which lessons learned in the field can be captured and recorded into doctrine, education, and training. Yet in each case, powerful political and cultural forces averse to small wars and occupation managed to sideline the efforts to institutionalize lessons for future generations. Thus, although this period yields clear evidence of learning from experience and clear examples of how organizations can create processes for learning from experience, the Army and Marine Corps seemed to cope and adapt more than they actually learned institutionally.

Coping and Adapting in the Early Years

Contrary to current conventional wisdom, the initial purpose of the standing American Army was not simply to "fight and win the nation's wars" but, rather, to promote the development of the nation by "intimidating" and fighting the Native Americans and guarding against European colonial powers. These requirements were spelled out in *Sentiments on a Peace Establishment,* written by General George Washington in 1783. In the document, Washington outlined the way the new nation's military should be constructed and the purpose it would serve.

> A regular and standing force, for Garrisoning West Point and such other Posts upon our Northern, Western, and Southern Frontiers, as shall be deemed necessary to awe the Indians, protect our Trade, prevent the encroachment of our Neighbours of Canada and the Floridas, and guard us at least from surprizes.[3]

Accordingly, graduates of the United States Military Academy simultaneously comprised the Corps of Engineers and the regular Army. According to military historian Russell Weigley, President Jefferson, who founded the Academy, believed that "if a standing army must be tolerated . . . it should be as useful as possible, and not merely in military ways."[4] Thus, in addition to being trained as soldiers, West Point men were expected to "serve the nation as a citizen in peaceful pursuits."[5]

This emphasis on engineering prepared the Army well for its nation-building task during the 19th century. Following the poor performance of the regular Army in the War of 1812 however, the civil engineering focus of the curriculum gradually broadened to include more of the classic "military arts." Nineteenth-century American military professionals became eager students of their more mature counterparts across the Atlantic. The Napoleonic Wars had provided a veritable classroom for military theorists, such as Baron Antoine Henri de Jomini, author of *The Art of War* (1838), and Carl von Clausewitz, author of *On War* (1831).

Although career soldiers of the 19th century did fight "real" wars such as the War of 1812 and the American Civil War, most still spent the majority of their careers on the frontier.[6] This unpleasant duty involved what today would be best described as a combination of guerrilla war-

---

3. George Washington, "Sentiments on a Peace Establishment" (1783), http://www.potowmack.org/washsent.html.
4. Russell F. Weigley, *History of the United States Army* (New York: Macmillan, 1967), 105–6.
5. Ibid.
6. Andrew J. Birtle, *U.S. Army Counterinsurgency and Contingency Operations Doctrine 1865–1941* (Washington, DC: Center of Military History, 1998); John Gates, "The Alleged Isolation of U.S. Army Officers in the Late-19th Century," *Parameters: Journal of the U. S. Army War College* 10 (1980); John Gates, "Indians and Insurrectos: The U.S. Army's Experience with Insurgency," *Parameters* 13 (1983).

fare, nation building, peacekeeping, and even ethnic cleansing. However, due to the passions of West Point professionals, lessons learned during those missions would be virtually ignored in formal military doctrine and education, leaving officers to rely on little more than personal experience and shared stories to guide their actions in subsequent, "irregular" campaigns.

The Army's campaign against the Seminole Indians from 1817 to 1842 reflected the struggle to learn in the field. In the second and bloodiest campaign, from 1835 to 1842, a number of the Army's most famous generals attempted to defeat the Seminoles by bringing the enemy to the Army's orderly battlefield; but the Indians continued to wear down their adversary using tried-and-true guerrilla tactics. Gradually, the Army managed to adjust its approach, incorporating decentralized patrols that actively pursued the enemy.[7]

It was not until more brutal tactics were used in 1841 that the Seminoles were finally defeated. Instead of calling the Indian warriors out to fight, Colonel William J. Worth took the "fight" to the land itself. Commanding his men to fight through the brutal summer months, he destroyed the Seminole's homes and crops, literally starving the enemy and its people into submission. Despite public and congressional criticism of some of the harsh tactics eventually employed against the Seminoles (including the use of bloodhounds by General Taylor), the six-year resistance was ended by Worth in less than one year. Thus, one of the key lessons absorbed from this experience was how brutal such fighting must be in order to succeed. Similar tactics would be used in subsequent campaigns, with increasing moral controversy.

Once many of the Indian tribes had been defeated and relocated to reservations, frontier duty for America's soldiers entailed less warfighting and more nation building. Entire communities developed around some of the military forts in locations that are now major American cities. Military units spent most of their days building and maintaining forts and roads, growing crops, tending livestock, and acting as the lone source of law enforcement for settlers and Indians in their regions. During what are often referred to as the Army's "constabulary" years (most of the 19th century), troops acted as peacekeepers, preventing Indians and settlers from encroaching on each other's territory. As in peace operations today, military officers and their men proved well suited to the task of nation building and peacekeeping in semihostile and isolated environments, even as the learning curve was steep.

---

7. Gates, "Indians and Insurrectos."

Without established courts and judges, officers often negotiated disputes among tribes and among settlers. Historian John Gates paints a clear picture of how unclear the line often was between soldier and police officer.

> The Army's task was that of the police officer, to track down the guilty [Indians] and bring them back for punishment. Because of the numbers involved those activities sometimes looked like war, and in a few instances, when entire tribes rose up in arms to fight against the intrusion of the white, it was.[8]

Throughout this period, troops were formally prepared for traditional European warfare and then learned on the job for their "real" duties. If they were fortunate, they learned through semiformal lectures conducted ad hoc in the field by more experienced officers. "So long as the links in the chain of human memory remained unbroken," observed one historian, "the informal approach [to learning] sufficed."[9] In fact, given that formal doctrine as we know it today simply did not exist at the time, this informal approach was all that was available.

The lack of institutional systems for capturing field experience was the key impediment to organizational learning during this era. There was simply no formal means for transferring new ideas from the field to the institution as a whole. Although *Military Institutes,* the one volume on official military regulations, had introduced in 1821 the requirement for officers to submit written battle reports, such reports were used more for identifying outstanding soldiers for official recognition and awards than as an instrument of learning and improving tactics, techniques, and procedures. Accordingly, no lessons were captured in writing, and no official military doctrine on these tasks was written or taught.

Texts eventually written in the schools were based on theories and observation of European war, not on the American Army's own contemporary experience. The authors of such texts, such as Winfield Scott and Dennis Hart Mahan, had little or no personal experience on the frontier. Mahan's entire military career was spent in academia, and his formal training was as an engineer. Still, personal experience by officers serving in the West was a common reference for subsequent operations—including guerrilla operations and reconstruction duties in the American South and pacification in the Philippines and the Caribbean, indicating a generational, informal transfer of contemporaneous lessons learned.

---

8. Ibid.
9. Dennis J. Vetock, *Lessons Learned: A History of US Army Lesson Learning* (Carlisle Barracks, PA: U.S. Army Military History Institute, 1988).

Fighting Guerrillas in the Civil War

The Army's experience during the Civil War and Reconstruction required some of the same skills needed to fight Indians and keep the peace on the frontier. An organized Southern rebel insurgency and general banditry frustrated Union commanders from the start of the war, as did the military's politicized role in Reconstruction for the 12 years following the South's surrender. Unfortunately, except for some personal experience, many Union commanders found themselves and their troops woefully ill-trained and otherwise unprepared to face these tasks.

Previous "on-the-job training" substituted for the necessary doctrine, education, and training the Army failed to provide for irregular war. The lack of consistent doctrine or strategy for irregular operations meant that counterinsurgency and pacification tactics in occupied areas varied from region to region depending on who was in command. Union commanders and veterans of the Indian and Mexican wars such as General George Crook, Major General Sam Curtis, and Colonel Henry Lazelle shared the general sense that what had worked for them on the American frontier would work in Virginia and Tennessee against Confederate irregulars. Accordingly, familiar frontier counterguerrilla tactics such as the use of small-unit patrols, the implementation of pass systems and checkpoints, the raising of special counterguerrilla units from within the Army as well as from the local militia, and the use of local scouts and spies for rebel intelligence were applied.[10]

Commanders had also learned from previous conflicts that guerrillas relied on support from the population to fuel their operations. As General Halleck described it, they sought to address this behavior as they had on the frontier by making civilians "feel the presence of the war."[11] The goal was to motivate communities to withdraw support for the rebels and to convince civilians to give up insurgent fighters living among them. Again, personal experience of generals such as Halleck, Pope, and Wool guided the policies adopted by the Union forces. These included punishing entire communities with fines, banishing families, confiscating property, and arresting individual civilians suspected of aiding the insurgents. General Sherman adopted a policy of exiling 10 families every

---

10. Birtle, *U.S. Army Counterinsurgency and Contingency Operations Doctrine 1865–1941*; Jeffery D. Wert, *Mosby's Rangers: The True Adventures of the Most Famous Command of the Civil War* (New York: Simon and Schuster, 1990).

11. General Halleck quoted in Birtle, *U.S. Army Counterinsurgency and Contingency Operations Doctrine 1865–1941*, 30.

time insurgents on the Mississippi River shot at his boats. Civilians arrested were tried in military courts and, if convicted, were sometimes executed.

Although implementation of these counterinsurgent and civilian pacification tactics was generally understood by military commanders to be an effective approach to guerrilla warfare, President Lincoln bristled at the harsh tactics being used against the rebel insurgents and civilians. Sensitive to the eventual challenges of postwar reunification, the president continually attempted to moderate his field commanders' behavior, urging more lenient tactics. Over time, commanders convinced the president of the efficacy of harsh, but discriminate, tactics as the most effective way to quell insurgencies.[12]

Officers serving in the South were frustrated over the lack of guidance for how to treat civilians and irregulars. The debate led eventually to General Order 100, otherwise known as the "Lieber Code," issued on April 24, 1863, at the request of officers serving in the South. The order clarified who would and would not be considered a "combatant" and prescribed how irregular guerrillas and disloyal civilians were to be treated. The authors sympathized with Union commanders and, in Article 156, made it clear that civilians aiding the enemy could be dealt with harshly.

> The commander will throw the burden of the war, as much as lies within his power, on the disloyal citizens, of the revolted portion or province, subjecting them to a stricter police than the noncombatant enemies have to suffer in regular war; and if he deems it appropriate, . . . he may expel, transfer, imprison, or fine the revolted citizens who refuse to pledge themselves anew as citizens obedient to the law and loyal to the government.[13]

Thus we begin to see a blurred line between conventional and counterinsurgency warfare. Sherman's concept of "total war" in which he perceived the vulnerability and the strength of the South to be its citizenry seems a lesson learned from counterinsurgent warfare. Sherman was well known for his intolerance of civilians who aided guerrillas.

> In districts and neighborhoods where the army is unmolested, no destruction of property should be permitted; but should guerrillas or bush whackers molest our march, or should the inhabitants burn

---

12. Ibid.
13. General Order 100.

bridges, obstruct roads, or otherwise manifest local hostility, then army commanders should order and enforce a devastation more or less relentless, according to the measure of such hostility.[14]

Just as General Wool destroyed the crops and dwellings of the Seminoles in order to weaken the support structure of the warriors, so Sherman would eventually burn the cities, destroying all rebel infrastructure and demoralizing the population in his March to the Sea. Sherman's march, largely viewed as a turning point—however brutal—in the effort to bring the Civil War to an end, reflects the brutal efficacy of taking the war to the people. Despite inconsistencies and inefficiencies, the implementation of these harsh and hard-learned lessons from the Indian and Mexican wars enabled the Union Army to contain much of the insurgent rebellion in the occupied territory as well.

Historians and other interested scholars can debate the potential reasons why the South failed to launch a substantive postwar insurgency against the federal troops. Had a large-scale insurgency actually ensued, the Army might have been forced to think more seriously and institutionally about its counterinsurgency doctrine and tactics—and maybe about the link between the two types of wars. As it happened, however, no formal doctrine for this "other" Civil War was created, and no special tactics were codified and taught to future generations of officers and soldiers.

This lack of attention to the unconventional elements of the Civil War is understandable given the casualty rates for more conventional battles. Military units often sustained close to 50 percent casualties in a matter of hours in these bloody contests. In some of the more savage encounters, such as Antietam or Gettysburg, individual regiments lost over 80 percent of their soldiers. To better prepare for future conflicts, it is reasonable that military leaders would wish to focus their study on the most violent and large-scale phenomenon. In the end, the conventional war had offered military leaders the opportunity to put to test the European strategies and doctrines they had been studying and also validated the need to continue to focus on the same.

An Army of Occupation: Reconstructing the American South

Reconstruction forced the Army to take on national building and law-and-order tasks similar to those on the frontier, but on a greater scale and with greater political scrutiny. The provost marshals who had pro-

---

14. John Brinsfield, "The Military Ethics of General William T. Sherman," *Parameters* 12 (June 1982): 44.

vided legal structure in occupied territories during the war continued to act as the main source of governance in many districts. Despite their fluctuating and ambiguous legal status as occupiers and peacekeepers, generals such as Banks, Dix, Saxton, and others established small military governments to manage the crises in towns and on plantations from Washington, D.C., to Louisiana. By the time Congress finally established the War Department's Bureau of Refugees, Freedmen, and Abandoned Lands in 1865, Union officers, along with myriad ad hoc freedmen's aid societies, were already enfranchising freed slaves, monitoring elections, providing employment, managing payrolls, distributing confiscated lands, collecting taxes, settling grievances, and setting up school systems.

Resistance to the Army's presence, along with various political machinations during Reconstruction, resulted in a significant legal clarification of the military's role via the passage of the 1878 Posse Comitatus Act. This act, which is still in place today, subtly influenced the development of the professional military culture in the United States by reaffirming the separation between "military" and "police" duties. The purpose of *posse comitatus* was to end the practice of having federal troops monitor elections in the South. Because election monitoring was a fundamental aspect of the Freedmen's Bureau's strategy to lift freed slaves to equal citizenry, it was a clear irritant to the Southern elite. In addition to its challenging their status in society, white Southerners felt such federal oversight to be an intrusion on state sovereignty—and President Johnson agreed. Although the Southern elections were the primary motivation for the bill, the Army's actions on the frontier were also brought into question. Federal troops stationed on the frontier had become the only source of law enforcement in these regions, and there was a growing concern that some of their actions had become heavy-handed.

Although *posse comitatus* developed from Reconstruction-era politics, its effect has continued to influence attitudes about military roles–especially with respect to stability and peace operations. The bill's clarification on the line between "police" duties and "military" duties became embedded over time in the psyche of the American national security establishment in general and in the culture of the American military in particular.[15] Although the original purpose of the law was to prevent the politicized use of U.S. military forces against U.S. citizens, the law came to be interpreted as a prohibition on the use of U.S. military forces

---

15. Bonnie Baker, "The Origins of the *Posse Comitatus*," *Air and Space Power Chronicles* (November 1999); Major Craig Trebilcock, USA, "The Myth of Posse Comitatus," *Journal of Homeland Security* (October 2000).

in "policelike" activities in general. This has resulted today in elaborate rules of engagement for U.S. forces assisting the Border Patrol, the Coast Guard, and the Drug Enforcement Agency in the so-called drug war. Likewise, as many of the tasks associated with peace operations, such as riot control and patrolling, seem more like police work than military combat, many Americans view peace operations as an improper use of U.S. military forces—even if such modern activity is not on U.S. soil or directed against U.S. citizens. Thus, *posse comitatus,* a law born of the federal occupation of the American South, has added one more layer to the arguments against using U.S. forces for stability operations today.

Posse comitatus reinforced the notion that the frustrating duties the military had taken on in the South were inappropriate for professional soldiers. Thus, despite the sixteen years of irregular warfare, occupation, military governance, peacekeeping, and humanitarian relief the Army experienced while fighting and reconstructing the South, the Army recorded no new doctrines for these duties. As Andrew Birtle explained, "officers regarded their experience as an aberration, so unique that it was unlikely to arise again."[16] Not only was this type of military duty considered unusual, if not inappropriate; for many of those who served, it was downright objectionable. Upon his transfer to the West after serving for years in the South, Lieutenant General Nelson A. Miles said, "It was a pleasure to be relieved of the anxieties and responsibilities of civil affairs, to hear nothing of the controversies incident to race prejudice, and to be once more engaged in strictly military duties."[17]

Ironically, as previously discussed, "military duties" on the frontier and in the South still never reflected the type of warfare being taught in the formal military schools. As discussed below, attempts to improve the learning of historical lessons were just beginning at Fort Leavenworth, but the upcoming war against Spain would have little in common with the Napoleonic land warfare being studied at West Point. The naval battle against the Spanish fleet was over quickly, leaving field commanders in the Philippines and the Caribbean once again to deal with stability and reconstruction duties in the untidy aftermath of formal victory.

Learning to Learn with Colonel Arthur L. Wagner

The transformation of the collective experience of 19th-century nation builders and guerrilla fighters into lessons that had been learned and

---

16. Birtle, *U.S. Army Counterinsurgency and Contingency Operations Doctrine 1865–1941,* 57.
17. Ibid.

into institutional memory suffered not only from the obsession with European military theory and strategy but also from a lack of formal methods, systems, and organizational structures for gathering, analyzing, and disseminating information in general. The Army was like a small club, in which information sharing was swift as soldiers coped and adapted in the field. But this informal dialogue remained relatively isolated from the formal doctrine and education system. Most formal field reports were geared not toward learning from experience but for documenting heroic actions of soldiers. Reports were mostly filed and forgotten. Thus, lessons from more conventional military operations, such as the many battles of the Civil War, also suffered from too little contemporaneous analysis—and, perhaps as a consequence, too much bloodshed.

The lack of any formal structures for capturing and analyzing experiential knowledge meant that existing doctrinal guidance based on theory could not be updated contemporaneously to reflect new data or methods. Coping and adapting in the field simply failed to make up for a proper analytical system. Sharing information was necessary but not sufficient for true organizational learning. A formal system of collection and analysis might have revealed the effects of actions taken over time and allowed for the recognition of disparities between experience and theory. Due to the efforts of one man, Colonel Arthur L. Wagner, this organizational and procedural shortfall began to change in the latter half of the 19th century. Unfortunately, the maturation of the Army's system for lessons learning came too late for soldiers of the 19th century.

Colonel Wagner, a seasoned soldier with experience on the Western frontier, became an instructor at the Army's new professional military college for field-grade officers in Fort Leavenworth, Kansas, in the early 1880s. Wagner was the first to publish and teach from analytical assessments of past military operations.[18] His scientific approach included the review of heretofore-ignored written reports, complemented by personal observation, interviews with participants, and comparative analyses of like campaigns. For this, military historian Dennis Vetock identifies Wagner as a "transitional figure to modern lesson learning."[19]

Wagner's efforts corresponded to the post–Civil War American interest in the Prussian (vs. French) military style. Beyond strategy and tactics, American observers (many of whom traveled abroad to study the Prussian system) noted the Prussian processes for examining their past and

---

18. The Army's School of Application for Cavalry and Infantry, established at Fort Leavenworth, KS, in 1881, was the forerunner for the Army's Command and General Staff College.

19. Vetock, *Lessons Learned*.

learning from successes and mistakes. This system included planning and training for future wars based on the lessons gleaned from past experience—their own and that of other militaries.

Wagner and others sought to emulate the Prussian method of self-criticism and planning efficiency. Such systems play a key role in the ability of an institution to capture field experience and convert such data into useful doctrine, education, and training as institutional memory. Although Wagner's efforts still focused on "traditional" warfare more than irregular campaigns, his significant contribution was his introduction of the more analytical methods of inquiry in the development of doctrine, education, and training. For the soldiers of the 19th century, however, Wagner's efforts would founder in an immature organizational structure. Vetock explains,

> Wagner's reporting and analyzing of selected experiences foreshadowed the later role of training and doctrine observers, but, unlike his successors, he had no processing agency to send his information to or institutional procedures to transform it into doctrinal adjustments.[20]

The institutional structures for capturing and institutionalizing experiential lessons learned would continue to evolve through the Spanish-American War and into the interwar years. As will be discussed later in this chapter, the U.S. Army's continued trend toward Prussian-style professionalism, along with some of the processes put in place by Wagner at the turn of the century, meant an increased focus on formal historical learning as a part of a new, more systematic war-planning process prior to World War II. But these systems would develop too late to capture the experience of the Spanish-American War. Soldiers would cope and adapt over time through new and improved processes for sharing experience contemporaneously, but as discussed below, more rigorous analysis and institutional learning was stymied by immature systems as well as political interference.

Coping and Adapting in the Philippines and Cuba

Nineteenth-century officers had learned to cope informally through ad hoc group instruction on frontier garrisons and a few basic texts. By the time Army troops faced the *insurrectos* in the Philippines, many of the tactical lessons had been passed to the next generation in this informal way. From an organizational learning perspective, informal networks and personal experiential learning provided a foundation, albeit a weak one,

---

20. Ibid., 29.

for the pacification operations following the swift victory in the Spanish-American War. Soldiers in the Philippines and Cuba were also able to use new Army-sponsored journal articles to share experience. As discussed below, this instrument became a key vehicle for socializing the idea of "lessons learned" over time.

Formally, the basic texts to which soldiers were exposed at the time of the Philippine and Cuban campaigns provided a paucity of useful information. Writing of his personal Philippine experience in 1900, Lieutenant Colonel James Parker observed, "The methods of deploying laid down in the drill regulations were not always found applicable."[21] Two categories of formal publications—field regulations and laws of war—would have been taught in American military schools and widely referenced at the time (table 1).

For moral and legal guidance, General Order 100, first published during the Civil War, was republished in 1898 when many of the same issues regarding civilians aiding guerrilla forces arose in the Philippines. For enemy forces and civilians who flouted the laws of war, General Order 100 allowed for the destruction of villages, exiling of citizens, and other punitive measures used during the Civil War, but it did not recommend what should be done to achieve best results. Thus, the accepted principles of international law as presented in contemporary texts such as Halleck's *International Law* and others provided a moral framework for officers' actions (however questionable) but gave little in the way of con-

TABLE 1. Resources for Officers in 1898–1901

Legal References and Texts	Military Manuals
International Law (Henry Halleck, 1861)	*Troops in Campaign: Regulations for the Army of the United States* (1892)
Introduction to the Study of International Law (Theodore Woolsey, 1864)	*The Service of Security and Information* (Arthur Wagner, 1896, 1899)
Outlines of International Law (George Davis, 1888)	*Infantry and Drill Regulations* (1898)
Military Government and Martial Law (William E. Birkhimer, 1892)	*Organization and Tactics* (Arthur Wagner, 1901)
General Orders 100 (aka "Lieber Code," 1863, 1898)	*Cavalry Drill Regulations* (1892)

Note: The following authors have written extensive reviews of these manuals: Keith B. Bickel, *Mars Learning: The Marine Corps Development of Small Wars Doctrine, 1915–1940* (Boulder: Westview Press, 2001); Andrew J. Birtle, *U.S. Army Counterinsurgency and Contingency Operations Doctrine 1865–1941* (Washington, DC: Center of Military History, 1998).

---

21. Lieutenant Colonel James Parker, USA, "Some Random Notes on the Fighting in the Philippines," *Journal of the Military Service of the United States* 27 (June 1900).

crete doctrine or tactics. For more tactically oriented guidance, officers would have referenced Army manuals such as the 1892 edition of *Troops in Campaign* or *Infantry and Drill Regulations,* published in 1898—none of which contained any practical guidance for conducting small wars, nation building, or pacification.

Some of the best sources of guidance for officers in the field were the new military journals from the various branch schools (i.e., the Cavalry, Artillery, and Infantry). These journals actively sought articles from troops and thus acted as a conduit for sharing experience and lessons learned. Eventually, the term *lessons* and the idea of learning from contemporaneous experience began to appear in these journals. Essay contests for these journals were expressly designed to learn from the "lessons of the past." As discussed below, the Marine Corps would follow this practice, soliciting articles from those in the field during their two decades in the Caribbean and actually using journal articles as a basis for writing new doctrine.

In addition to the new journals, most officers in the Philippines relied on personal experience as their guide. Twenty-six out of the 30 generals who served in the Philippines had spent a good portion of their careers performing nation-building operations and battling irregular forces in the American South and on the Western frontier.[22] Officers such as General Chaffee, who had chased the infamous Southern partisan group Mosby's Rangers during the Civil War, and other officers who learned the utility of destruction from General Sherman's March to the Sea applied these same approaches against the *insurrectos* and the civilians suspected of supporting them in the Philippines. After four years of trial and error in the Philippines, a handful of familiar tactics had been rediscovered and implemented—with varying degrees of success.

For many junior officers with little or no prior frontier or occupation experience, however, reconstruction and governance were frustrating endeavors. Lacking doctrine, training, and resources—especially in remote areas where civilian administrators were scarce—these officers struggled to manage local governments and implement other nation-building tasks. These younger officers learned by trial and error how to balance "attraction" and "chastisement" policies. Eventually, as the insurgency grew—and in areas where it was strongest—carrots often gave

---

22. Max Boot, *The Savage Wars of Peace: Small Wars and the Rise of American Power* (New York: Basic Books, 2002).

way to sticks, and commanders focused on the more brutal aspects of guerrilla warfare.

Over time, the military's tactics to "crush" the *insurrectos* became increasingly forceful—and controversial. Tactics varied, as local commanders in different areas tried out different approaches. In an effort to isolate guerrillas from the population, villages were often garrisoned and burned, populations were placed in concentration camps, crops were destroyed, and torture—such as the notorious "water cure"—was used to extract information.[23] Following Sherman's example of banishing families who assisted rebels, a handful of guerrilla leaders and their families were even exiled to Guam. The use of indigenous forces, such as the Philippine Macabebe Scouts, the Philippine Constabulary, and other native units, supplemented American forces, provided useful intelligence, and could often be counted on to administer some of the harsher methods in the field. Just as the Army had capitalized on tribal rivalries on the Western frontier to gain the tactical advantage, so they played on longstanding hatreds within Philippine society to break down the resistance. Officers would find themselves defending the use of these hard-learned but brutal tactics, despite their apparent tactical utility.

When news of harsh tactics and various atrocities committed by American troops reached the United States, the American public was mortified. This was the sort of behavior they expected from cruel European colonists, not "benevolent" American liberators. Veterans of the Philippine War found themselves defending their counterguerrilla tactics. Officers serving in the Philippines felt that only when harsh tactics had broken the will of the insurgents would the population respond to the "policy of attraction."[24] Major General Loyd Wheaton wrote in 1900, "You can't put down a rebellion by throwing confetti and sprinkling perfumery."[25] Whether or not this is the case is open for analysis. The military's simultaneous mission to Cuba, where such harsh measures were not implemented—or perhaps where they did not need to be implemented—provides additional insight into this counterinsurgency puzzle.

---

23. In the water cure a captive was made to drink copious amounts of water, after which pressure would be applied to the painfully distended abdomen.
24. Boot, *The Savage Wars of Peace;* David Haward Bain, *Sitting in Darkness: Americans in the Philippines* (Boston: Penguin, 1986); Gates, "Indians and Insurrectos."
25. Quoted in Birtle, *U.S. Army Counterinsurgency and Contingency Operations Doctrine 1865–1941,* 135; and Stanley Karnow, *In Our Image: America's Empire in the Philippines* (New York: Ballantine Books, 1989), 179.

Cuba: Lessons Applied

In contrast to what occurred in the Philippines, a bloody Cuban insurgency was somehow avoided in the immediate aftermath of the Spanish-American War. There are a number of potential reasons for this. First, the three years of fierce guerrilla warfare by the Cubans against the Spanish, combined with the brief American invasion, had laid waste to the island and its infrastructure and likely exhausted the Cuban people. Second, the fact that the Teller Amendment made it clear to the Cubans that the United States had no desire to colonize the island may have quelled potential animosity aimed at would-be occupiers. This concept was reinforced by General Brooke's actions upon assuming control of the island in January 1899. Over the course of the year, General Brooke established indigenous provincial governments under the tutelage of his own officers, thus reinforcing the concept of local rule while ensuring a level of competence as the new officials became accustomed to their roles. Finally, Brooke's decision to immediately demobilize the 50,000 men in the Cuban Army of Liberation by paying the soldiers, buying their weapons, and providing retraining for employment as policemen was probably the single most effective means of quelling any potential rebellion. The havoc an abruptly unemployed band of young men with weapons can wreak on an occupying army would become all too apparent 100 years later, following the fall of Saddam Hussein in Iraq.

On the Americans' second tour to Cuba, in 1906, significant violence was also averted—this time by veteran officers of the Philippine insurrection who comprised the command and staff of President Roosevelt's cadre. Anticipating guerrilla warfare, Marine Corps colonel Littleton Waller and Army general Bell applied some of the hard-learned lessons from their Philippine experience.[26] They immediately ordered troops to garrison towns, conduct patrols, and collect intelligence on potential rebel sympathizers. Arguably, these measures proved effective in preventing would-be guerrillas from conducting their operations.[27]

Meanwhile, on the nation-building side, Cuba was a "workshop for American progressivism."[28] General Leonard Wood (after whom the U.S. Army engineers' school is named) was praised by fellow Progressives for his aggressive program of nation building. They considered the gen-

---

26. See Boot, *The Savage Wars of Peace*, 120.
27. Keith B. Bickel, *Mars Learning: The Marine Corps Development of Small Wars Doctrine, 1915–1940* (Boulder: Westview Press, 2001), 28–42.
28. Gates, "The Alleged Isolation of U.S. Army Officers in the Late-19th Century."

eral "the very embodiment of a modern major general, one who could not only destroy nations, but build them."[29] This predilection for such tasks might seem incongruent with the traditional military aversion to such duties. But as historian John Gates points out, the Progressive movement, which was taking hold in the United States at the time and leading many influential Americans to believe in the ability of humankind to make the world a better place, likely influenced the thinking and actions of the era's professional military officers.

Although some scholars, most notably Samuel Huntington, have argued that isolation on the frontier set officers apart from contemporary intellectual thought, Gates convincingly rejects this theory.[30] He points out that most regular officers at the time had graduated from West Point. Because of the rigorous academic admissions standards and the fact that an appointment to the Academy required congressional sponsorship, this meant that West Point men were preselected members of the elite class of American society. Furthermore, "isolation" on the frontier was punctuated by long sabbaticals back to "civilization," during which time many officers pursued advanced degrees and taught at civilian institutions. Given this perspective, the enthusiasm and competence with which many of these officers pursued occupation and pacification duty during this period of American history is less surprising. Moreover, the engineering skills many acquired at West Point complemented their political predilections and enabled them to design and direct countless public works projects.

Despite the enthusiasm among this generation of "progressive" officers, lessons from this era were still not recorded into formal doctrine. The handful of military manuals that underwent periodic revisions in the years following these campaigns hardly reflected any pacification experience at all. *Troops in Campaign* was substantively unchanged in its 1903 revision, as were Wagner's *The Service of Security and Information* and *Organization and Tactics* in postwar editions. The *1905 Field Service Regulations* discussed guerrilla warfare in the context of international legal and moral obligations for the treatment of disloyal citizenry but offered nothing in the way of useful tactics. The two exceptions were the *1911 Infantry and Drill Regulations* and the *1914 Cavalry Drill Regulations*, both of which identified "minor wars" as a distinct category of warfare. These

---

29. Birtle, *U.S. Army Counterinsurgency and Contingency Operations Doctrine 1865–1941*, 106.

30. Samuel Huntington, *The Soldier and the State: The Theory and Politics of Civil-Military Relations*, 18th ed. (Cambridge, MA: Belknap Press of Harvard University Press, 2001).

regulations presented an overview of the nature and tactics of insurgency but offered only limited tips on countertactics. Most of the successfully implemented tactics from the Philippine and Cuba campaigns, such as garrisoning, intelligence collection, search and destroy, and civic actions in general, were simply not discussed. In his analysis of these manuals, Keith Bickel identifies these small efforts in the *1914 Cavalry Drill Regulations* as the "high water point in codifying formal counterinsurgency doctrine." After that time, he claims, "all discussion of guerrilla warfare was removed from subsequent editions."[31] Any potentially useful lessons learned on these issues remained confined to the memories of those who were there.

This failure to capture lessons learned probably had as much to do with political guidance from Washington as it did with the immaturity of the emerging institutional structures for capturing lessons. By the early 20th century, the expectation that the Army would no longer be used for "small wars" missions was buttressed when Elihu Root, secretary of war from 1899 to 1904, proclaimed that the "real object of having an Army is to prepare for war." Later, Secretary of War Henry Stimson agreed with his predecessor, claiming that use of the Army abroad was more likely to be construed as an act of war, while marines could be deployed without giving the same impression.[32] Thus it became policy that small wars were to be conducted by the nation's more expeditionary service, the U.S. Marine Corps, leaving the Army to focus on preparing for "real" war.

The Army, exhausted by the complexities and uncertainties of guerrilla warfare and eager to shed the negative publicity regarding the atrocities committed in the Philippines, was more than happy to relinquish the job to its sister service. Political sensitivities over reported atrocities committed—especially in the Philippines—were enormous. Congressional inquiries and courts-martial had already been convened. Any honest analytic investigation of the Army experience would necessitate a full accounting of methods used, to include the various atrocities that had been committed. That the Army leadership was eager to avoid such self-examination was clear by the fact that the one officially sanctioned Army study of the campaign, Captain J. R. M. Taylor's *The Philippine Insurrection against the United States,* was never released. Thus, Wagner's methods for observation and lessons learning would increasingly be applied by the

---

31. Bickel, *Mars Learning,* 49–50.
32. Allan R. Millett, *Semper Fidelis: The History of the United States Marine Corps* (New York: Free Press, 1991).

Army to conventional military arts; but for small wars and counterinsurgency, the Army's 19th-century experience would not be formally captured and analyzed for future generations. As a result, when the Marine Corps began pacification duties in the so-called Banana Wars 10 years later, it would have to relearn many of the same lessons.

The Marine Corps Learns to Learn

From 1914 to 1934, the U.S. Marine Corps found itself conducting pacification duties (a.k.a. "small wars") in Haiti, the Dominican Republic, and Nicaragua. In these so-called Banana Wars, the Marine Corps' operational tasks were remarkably similar to those with which the Army had struggled during the previous century. However, in contrast to the Army, which failed to record useful formal doctrine reflecting the lessons learned in these missions, the Marine Corps' experience resulted in the publication of the *Small Wars Manual*. Keith Bickel's account of how actual Banana Wars veterans wrote this manual during their tours as instructors at the Marine Corps school in Quantico provides excellent insight into how the Marine Corps learning system evolved. This achievement reflected the growing professionalism of the Corps and its internal struggle for a distinctive mission.

The Marine Corps' creation of the *Small Wars Manual* in the 1930s and the Army's establishment of the civil affairs and military government programs in the interwar years (discussed below) are the first systematic examples of institutional military learning. Both examples reveal the ability of military organizations to learn from the past via the implementation of formal processes versus simply coping and adapting contemporaneously. On the other hand, each also reveals the dampening effect of political pressures on military learning for nontraditional military missions.

In the case of the Marine Corps, advocates of small wars as a core Marine Corps mission were confronted by strong political actors, both military and civilian, who felt such missions were an improper use of Marine Corps talent. Likewise, strong military and civilian political sentiment against using soldiers as military governors in occupation duties acted against Army efforts to study and train for civil affairs and military governance. The interplay between the steadily improving internal mechanisms for military learning and the oscillating political pressures regarding the proper application of military force is a perennial theme of American military learning.

Marines Corps Tactics, Techniques, and Politics in the Banana Wars

On July 28, 1915, 330 U.S. marines landed in Port-au-Prince, Haiti. This expedition, commanded by the infamous Philippine War veteran Colonel Littleton Waller, marked the beginning of the Marine Corps' Caribbean exploits in the so-called Banana Wars. From 1915 to 1934, they fought insurgents, pacified populations, and otherwise occupied and governed Haiti (1915–34), the Dominican Republic (1916–24), and Nicaragua (1927–33). They performed these missions—as the Army had before them in Cuba and the Philippines—without the benefit of formal doctrine, education, and training. Unlike the Army, however, who failed to capture lessons from its experience, the Marine Corps emerged from its 20-year episode in the Caribbean with a new cache of formal doctrine, informal materials on lessons learned, and new courses specifically tailored to the unique requirements of small wars.

The personal experience of a few seasoned officers, trial and error, and on-the-job training led the Marine Corps to initiate (eventually) many of the same tactics and techniques that the Army had used in the Philippines and Cuba. These included garrisoning towns, small-unit patrols, night operations, and the development of local constabularies. At first glance, it appears that Army lessons had somehow been transmitted to these marines. However, as scholar of small wars Keith Bickel points out, most of these tactics were not implemented quickly and appear to have been inefficiently rediscovered as the mission progressed: "Only after months of this constant, growing harassment did the marines switch to the operational offensive and systematically begin developing better intelligence measures, creating a constabulary, and using combat patrols."[33]

Although there are no archived records (personal letters or reports) indicating what these commanders were actually thinking, there are two plausible explanations for this apparent failure. First, compared to previous Army operations, marines were sent to the Caribbean under significantly different political circumstances.[34] Army operations in the Philippines and Cuba took place in the immediate aftermath of a "major war," where, according to the *Small Wars Manual*, "'diplomatic' relations are summarily severed at the beginning of the struggle." Marines operated in the Caribbean under different political expectations. In each of the three Caribbean operations, marines were not officially "at war"—

---

33. Bickel, *Mars Learning*, 75.
34. Lester Langley, *The Banana Wars: United States Intervention in the Caribbean, 1898–1934* (Lexington: University Press of Kentucky, 1985).

nor had they just fought and won a war. This distinction is addressed in the opening paragraphs of the *Small Wars Manual:* "As applied to the United States, small wars are operations undertaken under executive authority, wherein military force is combined with diplomatic pressure in the internal or external affairs of another state."[35] Thus, diplomatic sensitivities and State Department expectations reinforced the notion that the marines were to operate in a more bounded environment.

These political realities reinforced the second potential reason for the Marine Corps' failure to apply more aggressive tactics at the beginning of the operation: the Corps' own institutional history. The majority of Marine Corps land operations prior to 1915 had been very short, rather uncomplicated affairs. Thus, the limited political objectives in the Caribbean (to "restore order" or to "protect American interests") likely led them to assume upon landing that these missions were more likely to mirror their own short expeditionary landing operations of the past rather than the longer, more complex operations the Army had experienced. As one historian points out, "the marines' experience was in taking towns, not running them or negotiating with the local bigwig."[36]

These differences may have clouded the marines' abilities to note the more operational or tactical parallels to previous operations conducted by the Army, even despite the personal experience of a handful of Marine Corps officers and men in those operations. This highlights a recurring ironic theme about small wars—the tendency for military thinkers to view each operation as a sui generis event. In any case, to the extent marine commanders in the Caribbean were aware of the potentially useful tactical details of the Army's prior experience, it seems clear that marines did not necessarily view that experience as relevant to their current strategic situation.

As the intervention wore on, the Marine Corps needed to adapt. Speaking of his experience in the Caribbean, Lieutenant General Edward Craig stated, "We received no training when we were ordered to these places."[37] This was especially true for marines landing in Haiti in 1915. By the time others landed in Nicaragua in 1927, however, lectures on the topic had been added to the officers' courses at Quantico; a number of participants had received on-the-job and formal training in Haiti, the Dominican Republic, or both; and the Corps had begun distributing

---

35. USMC, *Small Wars Manual of the United States Marine Corps* (1940; repr., Washington, DC: Government Printing Office, 1987).

36. Langley, *Banana Wars.*

37. Lieutenant General Edward Craig, "Oral History," 30, quoted in Bickel, *Mars Learning,* 144.

"lessons learned" straight to the field in the form of journal articles in the *Marine Corps Gazette*. Although most of these efforts were on the margins of the formal Marine Corps system, they were a start. Significantly, they reflected an institutional learning process that had begun with the Army in prior campaigns but, as discussed previously, had not matured before the end of the Army's campaigns in the Philippines or Cuba.

The curriculum at the formal Marine Corps schools lagged behind operational experience. Marines sent to Haiti and the Dominican Republic had been schooled in traditional warfare in the courses for company and field officers. It was not until 1924, when the field-grade officers' school added seven mandatory 50-minute lectures on the topic, that officers had any formal academic exposure to small wars. Even this was a marginal effort, given that the full course contained a total of 750 hours of instruction. A course dedicated to small wars was developed the following year, but still these schools' emphasis on small wars was less than 1 percent of the overall curriculum.[38] In sum, officers in the field could not rely on the formal education and training provided by the Corps to prepare them for their mission.

To make up for some of this deficit, marines attempted to train onsite. In Haiti, there were schools for officers and noncommissioned officers (NCOs) to study basic military skills, but they were not specifically tailored to the problems of small wars, nor were they particularly well respected. There were attempts at brigade and regimental training courses as well in both Haiti and the Dominican Republic, but these, too, focused on the basics and could not provide a source of training specific to small wars. Additionally, during this period, a number of exercises designed to test the evolving doctrine of amphibious warfare were conducted on Caribbean islands. The development of this doctrine was an enormous priority for the Marine Corps leadership in the interwar years and a source of contention for advocates of a Corps focus on small wars.[39]

The notable exception to this poor training was the school for the Haitian Gendarmerie formed by Colonel Smedley Butler, which explicitly emphasized the various civil tasks required of officers in these operations.[40] Because the Gendarmerie officers were American, this became a

---

38. Ibid.
39. Allen S. Ford, "The Small Wars Manual and Marine Corps Military Operations Other than War Doctrine," in *Fort Leavenworth Papers* (Ft. Leavenworth, KS, 2003); Victor H. Krulak, *First to Fight: An Inside Look at the Marine Corps* (Annapolis: U.S. Naval Institute, 1984); Millett, *Semper Fidelis;* Langley, *Banana Wars.*
40. Bickel, *Mars Learning.*

source of formal learning and dissemination of knowledge for these unique tasks. Colonel Butler even issued his own pamphlet, *Gendarmerie Rules,* which provided additional guidance. Unfortunately, marines arriving in the Dominican Republic in 1916 would have no exposure to this training. In the Dominican Republic, it was three years before officers opened a training center relevant to small wars. This center allowed companies to rotate for six weeks of training in the full range of civil and military occupation duties. Later, in Nicaragua, training was accomplished at the unit level by individual commanders—many of whom had received little or no training in "bush" warfare themselves.[41]

Also attempting to compensate for the lack of formal education and training were volumes of articles, papers, and pamphlets that were written by experienced (mostly field-grade) officers during this period and became increasingly available to officers in the field. The greatest, most widely available source for these lessons was the *Marine Corps Gazette.* Because every marine had a subscription to the *Gazette,* it is reasonable to assume that concepts presented in these articles had fairly wide distribution. For officers serving in Haiti and the Dominican Republic, 13 out of 50 *Gazette* articles (less than 4 percent) published between 1918 and 1926 covered topics relevant to small wars. Most of these were published in the later years, leaving the first marines to land in Haiti at a distinct disadvantage. In contrast, between 1927 and 1941, a total of 34 new articles on small wars were published. Moreover, like the journal articles distributed by the Army in the Philippines and Cuba, these articles held a semiofficial status, having been specifically solicited for publication by the Marine Corps headquarters in Quantico.

The dissemination of these articles was valuable but still no substitute for well-reasoned doctrine and education. Most articles were personal histories and observations or unfiltered reports from units in the field. As such, they offered little in the way of well-analyzed guidance or synthesis. The few exceptions, such as the works by Major Harold H. Utley and Major E. H. Ellis, approached the topic more holistically. These authors expanded on their personal experience with literature from other sources. One commonly cited work was *Small Wars: Their Principles and Practice,* by British colonel C. E. Callwell. Callwell's work was published in 1896 and reflected the British lessons of colonial warfare. That marines were tapping into such work reflected their realization that their missions in the Caribbean echoed common experiences from which more formal doctrine could be derived. Thus, these articles reflected a con-

---

41. Ibid., 138, 95.

certed effort by a number of officers to comprehend the nature of these operations and to provide better guidance. Their efforts formed the basis for the *Small Wars Manual*.

The *Small Wars Manual*

The Marine Corps' *Small Wars Manual* may have been the first serious attempt in the American military to capture formally the lessons of small wars and to translate the lessons into official doctrine. There are references to a 19th-century Army handbook, *Small Wars and Punitive Expeditions,* but historians have been unable to locate copies in Army archives or to determine the degree to which the work influenced operations in the Banana Wars or the drafting of the *Small Wars Manual* during the 1930s.[42] To facilitate the writing of the *Small Wars Manual* in 1935, the new commandant of Marine Corps schools, Brigadier General James Breckenridge, ordered that all instruction on small wars be suspended at the Marine Corps schools. Instead of teaching their course that academic year, the instructors were to write the first Marine Corps formal doctrine for small wars. The product was a series of pamphlets that became the basis for the *Small Wars Manual* published in 1940. Unfortunately, the new doctrine was used in the schools for only 4 years (1936–40) before attention became fully focused on the problems of amphibious landing operations for World War II. By 1941, courses on small wars were simply no longer taught.

The *Small Wars Manual* was based on work previously published as *Marine Corps Gazette* articles by officers such as Utley, Harrison, and Ellis. Mirroring the method of those previous articles, authors writing on small wars referenced personal experience as well as Army doctrine and foreign work to cull the "best practices" for small wars operations at the time. Led by Major Utley, the instructor-writers also tapped the personal experience and insight of school students who were encouraged to participate in the process and review drafts. In this way, these authors adopted the techniques of Colonel Arthur Wagner, who had utilized past reports, interviews of participants, and comparative analysis to produce texts and lectures for the Army's Command and General Staff College. The final Marine Corps product was a 428-page work presenting detailed tactical-level guidance on logistics, administration, and combat operations for small wars, including such topics as infantry patrols, river crossings, animal care, and military government.

---

42. Correspondence with Keith Bickel and Conrad Crane.

Despite the fact that marines did conduct or direct the labors of locals in many civil works projects, there is no mention of roles such as road or school building, education, or other public works projects—all serious elements of the Army's progressive programs in Cuba and the Philippines. The emphasis in the manual is clearly on restoring law and order, not necessarily on reconstruction or on long-term improvements to the societies and countries to which the marines were sent. Thus, in contrast to the experience of the Army during the previous campaigns—and even to the experience of the Marine Corps today—the discussion of civil affairs centers around the *supervision* of infrastructure, not the *developing* or *building* of it. For example, paragraph 13.16 states,

> Public Utilities.—a. Municipal water works, light and power plants should be permitted to remain open and function as in normal times, but should be supervised by the officer on the civil affairs staff having jurisdiction over public works . . . Railways, bus lines, and other public carriers needed for military purposes may be seized and operated by the public works officer of the military governor's staff . . . This officer is also responsible for the upkeep of the highways.

This language reflects a fundamental, if overlooked, distinction between the small wars experiences of the Army and the Marine Corps. The *Small Wars Manual* reflects the political constraints under which marines operated in Haiti, the Dominican Republic, and Nicaragua, not the broader, more progressive mandate of the Army after the Spanish-American War. Unlike the Army in the Philippines and Cuba, who operated transitional governments in postconflict, reconstruction and occupation scenarios, the marines were deployed as a stabilizing force to countries whose governments, however fledgling or fragile, were still nominally—and legally, as far as the U.S. State Department was concerned—in control. Had the Army produced a similar doctrinal compendium following their operations in the former Spanish colonies, a greater emphasis might have been made on the utility (or futility) of reconstruction tasks such as building schools and roads or other public works and infrastructure projects and on helping a country develop its governing institutions.

Although a few Army officers were consulted in the writing of the manual, the final product very much reflects the Marine Corps' experience in the Caribbean prior to World War II. As such, it fulfills one of the goals of Marine Corps doctrine writers of the time—to differentiate Marine Corps doctrine from that of the Army and the Navy. Close examination of the manual suggests that had the doctrine been developed to encompass the combined experience of the Army *and* the Marine Corps from the Indian Wars, Southern Reconstruction, and the Philippines to Haiti and Nicaragua, it may have looked a bit different. Specifically, had a true "joint" approach been taken to writing small wars doctrine at that time, the final Army-Marine manual might have reflected a greater emphasis on the potential civil tasks officers might be required to accomplish. Instead, in sharp contrast to the level of detail provided for "laying ambushes," "attacking houses and small bivouacs," "river crossings," and other such purely "military" operations (each warranting multiple pages of discussion), the *Small Wars Manual* makes only passing references to public works or civil transportation.

Political Resistance to Learning

Contrary to conventional wisdom today, the Marine Corps was as far from embracing the small wars mission as its core duty as the Army was before it, if not more so. Although most marines during this era spent the greater portion of their military careers in the Caribbean, publication of the *Small Wars Manual* in 1940 was by no means foreordained. The manual's authors were not in the mainstream of the Corps' officership, nor were they aligned with the sentiments of the more powerful Marine Corps leadership of the day. Indeed, the project was driven by a few dedicated and experienced Marine Corps officers who believed that, for better or worse, the small wars role was likely to be a perennial one for the Corps.

In the 1920s, Corps commandant General John A. Lejeune actively promoted the development of the Quantico schools to encourage the study of amphibious warfare, not small wars; and his four successors, from 1929 to the start of World War II, all shared this sentiment.[43] Even as Banana Wars veterans were working to finalize their first draft of the *Small Wars Manual,* General John Russell (commandant from 1934 to 1936) claimed that "expeditionary forces in 'small wars' were not in the Corps' nor the nation's interests."[44] This enthusiasm by senior Marine Corps leadership for amphibious warfare reflected the drive to define a new and unique Marine Corps mission and was in sharp contrast to the assertion by the civilian leadership, as discussed above, who felt that the Marine Corps, not the Army, should support such small wars operations.

---

43. Millett, *Semper Fidelis.*
44. Allan R. Millett and Peter Maslowski, *For the Common Defense: A Military History of the United States of America* (New York: Free Press, 1994).

For many Marine Corps historians, the emphasis on amphibious warfare over small wars missions was validated in light of the role the Marine Corps played in World War II. Although the Army conducted many more amphibious landings during that war, the Marine Corps is rightly credited for having developed the doctrine. Still, by 1958, over 6,000 marines were landed in Lebanon, where they remained (along with 8,500 Army troops) as a stabilization force for 102 days. In 1963, another 8,000 marines found themselves in the Domincan Republic, in what Millett calls a "striking repetition of its duties during the colonial infantry era." The follow-on decade of counterinsurgency warfare conducted in Vietnam makes such duties seem far from a "transitory" distraction. Indeed, small wars were in many ways conducted in a nearly continuous stream from the end of World War II through Vietnam and beyond.

It should also not be overlooked that in each of these post–World War II missions, the Marine Corps was not operating alone. Although marines were usually the first to land, they were also the first to leave. Having had a respite from small wars duties prior to World War II, the Corps' sister service, the U.S. Army, was now responsible for land operations of longer duration. This division of labor reflects the Army's experience in longer-term occupation duties—duties it attempted to perfect following World War I. As discussed below, thanks to the newfound emphasis on historical review and war planning, the Army developed a program for "occupation duties," including military governance and civil affairs, in the interwar years. Like the Banana Wars marines, these Army advocates of military governance programs fought powerful internal and external political forces in their pursuit of change.

Frustrations Identified in World War I

Writing of his experience as the American officer in charge of civil affairs at the end of World War I, an exasperated Colonel Irvin L. Hunt exclaimed,

> It is extremely unfortunate that the qualifications necessary for a civil administration are not developed among officers in times of peace. The history of the United States offers an uninterrupted series of wars, which demanded as their aftermath, the exercise by its officers of civil governmental functions. Despite the precedents of military governments in Mexico, California, the Southern States, Cuba, Porto Rico, Panama, China, the Philippines and elsewhere, the lesson has seemingly not been learned. In none of the service-schools devoted to

the higher training of officers, has a single course on the nature and scope of military government been established. The majority of the regular officers were, as a consequence, ill-equipped to perform tasks differing so widely from their accustomed duties.[45]

Although coping in the field and relying on previous personal experience was nothing new for the American Army, the way in which the Army responded during and after World War I reflects the profound institutional changes that had begun with Colonel Arthur Wagner and continued during the war. The frustration of military commanders on the ground and the lessons gleaned from their experience were for the first time reflected in institutional and intellectual changes in the U.S. military.

As Dennis Vetock's comprehensive study of Army learning reveals, the first American attempt to institutionalize a modern contemporaneous lessons-learned system occurred during World War I. The rapid mobilization and overseas deployment of the U.S. Army meant that troops would need to be trained overseas. According to Vetock, "the birth of the Army's lessons learning occurred overseas because, basically, that was where the war was and where the AEF's [American Expeditionary Force] Commander-in-Chief, General John Joseph Pershing, functioned practically as a proconsul, fully supported—even deferred to—by the nation's civilian leadership."[46] Under the powerful leadership of Pershing, training and doctrine development took place in theater and was specifically designed to incorporate "changes suggested by actual experience."[47] As it happened, this experience included that of the war-weary allies as well as the experience of U.S. forces once they began to fight. The structure of the general headquarters (GHQ) and the system developed under Pershing's leadership facilitated this fledgling lessons-learned process and demonstrates the importance of institutions and top-down leadership in facilitating bottom-up learning.

In an effort to catch up to the allies, Pershing placed an enormous emphasis on training. A key factor in the formalization of the lessons-learned system was the establishment of the training section, led by the

---

45. Colonel Irwin L. Hunt, Officer in Charge of Civil Affairs, Third Army and American Forces in Germany, "American Military Government of Occupied Germany, 1918–1920" (March 4, 1920), 56–57 (aka "Hunt Report"), OCMH files, as excerpted in Harry L. Coles and Albert K. Weinberg, *Civil Affairs: Soldiers Become Governors, U.S. Army in World War Two* (Washington, DC: U.S. Army Center for Military History, 1992).

46. Vetock, *Lessons Learned*.

47. AEF, GHQ, *General Order No. 8* (July 1917), 9, quoted in Vetock, *Lessons Learned*, 39.

assistant chief of staff, G5 (training), under the GHQ. The G5 was responsible for training, doctrine, and "experience processing" and was authorized to adjust doctrine, as required, based on current experience. At first, training and doctrine manuals were updated based on French and British experience. Once American forces entered combat, however, the system switched to capturing and processing American lessons directly from the field. Foreshadowing the late 20th-century lessons-learned system, the G5 utilized both active observation and rapid dissemination procedures.

G5 "inspectors" (like modern-day "observer controllers" and "collection teams," discussed in chapter 4) roamed the battlefields and training centers, where they collected information through interviews and personal observations. As Vetock explains, unit battle reports lagged behind this shortcut system, as "oral observations, as well as written reports, became almost immediately available for processing . . . thanks to the pervasive use of the GHQ's 'inspector-instructors.'"[48] Their assessments fed directly into the training system through adjustments to training practices and publications.

The products of this system were training and doctrine lessons that were disseminated through orders, memos, training curricula, and manuals. The GHQ produced over 160 new publications during the period of the AEF's involvement overseas. Many of the new lessons identified were not just considered "good ideas" but became directive in nature through the use of *Combat Instructions,* with the memo acting as official military orders. The two editions of *Combat Instructions* were widely distributed (55,000 copies) and made clear to commanders that they would be responsible for understanding and implementing the new doctrinal information outlined therein. The second printing of *Combat Instructions* synthesized and distilled trends and other lessons identified in the more informal publications.[49]

More informally, lessons were disseminated in three major publications: *Remarks concerning Deficiencies in the Training of Our Units as Brought Out in Some of the Recent Offensive Operations, Report of American Officers on Recent Fighting,* and *Notes on Recent Operations.* Both *Remarks* and *Report* were onetime publications with a distribution of a few thousand, while *Notes* was printed after each major battle (four times in all) with a distribution of 15,000. Additionally, over 4,000 copies of a new training manual, based on previous experience, were distributed toward the beginning

---

48. Vetock, *Lessons Learned,* 47.
49. Ibid.

of the operation in an effort to disseminate the latest lessons learned from Allied experience. To facilitate this, Allied officers assisted in the training and doctrine development during the first stages of deployment.[50]

This system was credited by General Pershing for increasing combat proficiency and is an example of institutionalized contemporaneous lessons learning. The military did not simply permit the transfer of information as it had done previously, but it promoted learning by actively observing and disseminating lessons through systematic and official processes. The speed with which the system was implemented (the U.S. military was involved in actual combat operations for less than six months) makes it all the more impressive. It was, however, a wartime invention. Accordingly, it did not entirely survive the war's end: "Without active battlefields or, at least, major field maneuvers, the lesson-learning procedures of the G5 staff served no useful function."[51] Thus, after the war, the AEF G5 was dismantled, leaving the responsibility of capturing lessons from World War I to the historians and the planners.

As the following discussion of the interwar years reveals, while the AEF G5 did not survive after World War I, emphasis on the importance of capturing lessons and exploiting experiential learning was sustained. Even during the dramatic interwar downsizing, the roles and responsibilities of the arms of the Army responsible for historical analysis were further institutionalized. A newfound emphasis on examining the past and planning for the future took hold—culturally and organizationally—which enabled the Army for the first time to examine more formally the challenges of operations other than war (occupation, civil affairs, and military government).

Institutionalizing Civil Affairs and Military Government

Although there were no official doctrinal field manuals on civil affairs or military government published during the 1920s or 1930s, the topic had begun to generate interest among World War I veterans, some of whom were serving as faculty and students in the military's professional postgraduate schools. The small committee of military students studying "military government" at the Army's two professional postgraduate schools in the interwar period could not have envisioned the massive stabilization and reconstruction operation that was to take place in the af-

---

50. Ibid.
51. Ibid., 50.

termath of World War II.[52] Yet without the seeds planted by this small cadre of officers studying the Hunt report on World War I occupation in historical context, listening to briefings from World War I veterans, and preparing reports and pamphlets, officers in 1945 would have been no better off than their predecessors had been in the aftermath of that war. Before the first bomb had dropped on Pearl Harbor, steps had been taken, doctrinally and structurally, to ensure that lessons in stability and reconstruction would not have to be relearned in the field "next time."

Capturing lessons from the past during the interwar years was not an accidental outcome. According to historian Russell Weigley, "the duties assigned it by the 1920 National Defense Act (NDA) ensured that the Assistant Secretary's office would keep memories of the tremendous logistical problems of World War I alive."[53] Specifically, officers serving on the Planning Branch were ordered to review the mobilization records of World War I. This newfound emphasis on capturing lessons spilled over to the War College, where the Historical Section was compiling an official record of World War I and where Colonel Hunt and his colleagues were delivering briefings on the need to study military government and civil affairs.[54] Ironically, whereas the push from outside the institution in the form of the NDA catalyzed efforts by the Army to learn from its experience, outside pressures would also eventually bound their efforts by questioning the military's enthusiasm for military governance and civil affairs. Despite the fact that their examination of history revealed the perennial requirement for military forces to conduct such tasks, such roles smacked of imperialism to many civilian leaders.

In 1925, building on the Hunt report, the Lieber Code, and issues raised at the 1907 Hague Convention, officer-students at the Command and General Staff College in Fort Leavenworth, Kansas, produced a comprehensive manual, *Military Aid to the Civil Power*. This manual and a supplement that followed became the standard text on the topic and were used for decades. The approach at this stage was more legal and philosophical than doctrinal or tactical. Army historian Earl Ziemke explains, "The tendency of the War College in the 1920s was to look at civil affairs and military government entirely as they related to military law, the as-

---

52. Although the two terms were often used interchangeably, the term *military government* was distinguished from *civil affairs* in that "civil affairs was military government conducted on one's own or friendly territory, and military government was civil affairs conducted in enemy territory." Ibid.

53. Ibid.

54. Earl F. Ziemke, *The U.S. Army in the Occupation of Germany, 1944–1946*, Army Historical Series (Washington, DC: Center of Military History, U.S. Army, 1990).

sumption being that they were not much more than the functions of observing and enforcing law."[55] Still, as the following excerpt from *Military Aid* suggests, the responsibility of militaries to be prepared for their postwar tasks was made clear.

> International law recognizes that, having overthrown the pre-existing government and deprived the people of the protection that government afforded, it becomes not only the right, but the duty of the invader to give the vanquished people new government adequate to the protection of their personal and property rights.[56]

The focus on the legal and ethical issues surrounding warfare had gained increased momentum during World War I, when the judge advocate general's (JAG) office expanded to 426 officers. Prior to this, such duties were filled in a relatively ad hoc fashion and often by lawyers from civilian life. This institutionalization and growth of the JAG Corps influenced the growth of the civil affairs doctrine and education, as the JAG was charged with building a military police force, running internment camps inside the United States, and, eventually, writing doctrine for military government.

Meanwhile, at the Army War College, more senior officers were introduced to the topic via the G1 (personnel) War Plans Division planning course. Various materials used in this course and produced by its students were repeatedly proposed as doctrinal manuals for use during occupation duties (table 2). These texts, including the *Basic Manual for Military Government by United States Force, Military Law,* and *Administration of Civil Affairs in Occupied Alien Territory,* all addressed, in part or in full, the topic of postwar military tasks and responsibilities. Repeated requests to publish such material as a formal doctrinal manual on military government were initially resisted by the JAG, General Gullion, who felt that international law and (later) FM 27-10 *The Rules of Land Warfare,* published in October 1939, covered the topic sufficiently.[57]

G3 (operations) planners on the General Staff (G1), who were becoming increasingly concerned with events in Europe, disagreed with General Gullion. Having read the reports from World War I and having been lobbied by War College committees and G1 on the need for a manual on military government, G3 strongly urged the JAG to begin

---

55. Ibid.
56. As quoted in Stanley Sandler, *Glad to See Them Come, Sad to See Them Go: A History of U.S. Army Civil Affairs and Military Government* (S. Sandler, 1994).
57. Coles and Weinberg, *Civil Affairs.*

crafting a manual for use in postwar situations. The product, FM 27-5 *Military Government and Civil Affairs*,[58] was published in July 1940 (the same year the Marine Corps published its *Small Wars Manual*). Together, *The Rules of Land Warfare* and *Military Government and Civil Affairs* became known as the "Old and New Testaments of American Military Government," as they formed the basis for the American approach to postwar operations.[59]

Training for Civil Affairs: The School of Military Government

Once military professionals had collectively begun to recognize military governance as a critical aspect of postwar military operations, the next step was to begin a training program. As directed in FM 27-5, the per-

TABLE 2. Interwar Development of Civil Affairs Doctrine and Education

Date	Event	Notes
1920	Hunt Report	Based on World War I Rhineland experience
1925	Command and General Staff course texts	*Military Aid to the Civil Power*, written by students and faculty; becomes textbook, with supplement used for decades
1919–WWII	War College G1 War Plans Division course materials incorporating civil affairs topics	1919: *G1 Reference Data* 1934–35: *Basic Manual for Military Government by United States Force* 1938–39: *War Department Basic Manual Military Government*, vol. 3; *Military Law*, part 4; and *Administration of Civil Affairs in Occupied Alien Territory*
October 1939	Official doctrine FM 27-10 published	*FM 27-10, The Rules of Land Warfare*, published by Army Judge Advocate's Office
July 1940	Official doctrine FM 27-5 published	*FM 27-5, Military Government and Civil Affairs*, published by Army Judge Advocate's Office
1942	Reports from officers	Analysis from first U.S. graduates of new British School at St. John's College (Cummings and Thompson)
April 1942	School of Military Government	School opens at University of Virginia; Brigadier General Cornelius Wickersham, Commandant
January 1943	CATP	Civil Affairs Training Program established
April 1943	CAD (Civil Affairs Doctrine)	Major General Hilldring appointed commander of new Civil Affairs Division on General Staff

---

58. Departments of the Army and Navy, *FM 27-5 Army and Navy Manual of Military Government and Civil Affairs* (1943).

59. Ziemke, *The U.S. Army in the Occupation of Germany, 1944–1946*.

sonnel section of the General Staff became responsible for ensuring that the Army had adequately trained personnel for postwar duties. Unfortunately (although somewhat understandably), as the country mobilized for war and the Army focused on recruiting and training officers for the immediate fight, the last thing Army G3 (operations) planners had in mind was the postwar. By late 1941, only two American officers, Lieutenant Charles Thomson and Major Henry Cummings, had received any formal training on military government. As part of the intelligence branch working with the British military, the two had been invited to attend Great Britain's brand-new eight-week military government program at St. John's College in Cambridge. The reports submitted by Thomson and Cummings to G2 (intelligence) eventually found their way to G1 and others charged with crafting an American training program in civil affairs, thus ensuring that lessons from the British experience were incorporated into U.S. training.

Ironically, the JAG, General Gullion, who had originally resisted the task of writing doctrine for military government, became the champion of the American training program in military government. Having recently been assigned to serve as the new provost marshal general (PMG) in addition to his duties as JAG, General Gullion was charged with creating and training a new military police force. Like the JAG Corps, military police duties had heretofore been an ad hoc affair. When such an arrangement proved immediately inadequate for the increased responsibilities of running stateside internment camps for Japanese Americans, General Gullion was directed to start a school for military police. As the magnitude of the looming postwar responsibilities became increasingly clear to General Gullion and others in G1, they proposed to add to this program a school of military government. Over the objections of the G3 who were still focused on the immediate need for new officers and men to fight, Army chief of staff General George C. Marshall approved the school of military government in February 1942.

In April 1942, the first class of fifty officers began a four-month program at the University of Virginia's new School of Military Government. Using FM 27-10 and 27-5, along with the reports on the British school from Lieutenant Thomson and Major Cummings, the school's new commandant, General Wickersham, developed a curriculum that focused on the Army's previous experience with occupation and pacification (in the Philippines, Cuba, etc.), as well as more theoretical and practical aspects of military governance. Wickersham hired civilian scholars from top schools to teach courses on Italian, German, and Japanese culture and politics, which complemented the legal expertise of some of the military staff. Still, the initial staff was small, including three civilian and nine officer-instructors plus 25 additional staff members.

In addition to classroom lecture, students were given problems to solve. One of the first studies they accomplished was to determine how many civil affairs officers would be needed in postwar Europe and Japan. Their answer, 5,000 to 6,000, was clearly beyond the capacity of the new school. In response to this surprising finding, additional political wrangling by the G1, the JAG, and the school commandant eventually resulted in the creation of the Civil Affairs Training Program (CATP), which provided instruction at 10 other universities across the country. By the end of 1943, the Army had increased the number of graduates from 50 per class to over 2,000 every four months.

Training in CATP differed from that at the University of Virginia in two respects. First, the CATP graduates were expected to work more directly with the local population, while the Virginia graduates were to work as staff officers. Accordingly, CATP curriculum focused more on language and culture and less on problem solving, which was the focus of the Virginia program. Together, these two programs reflected profound institutional and intellectual changes in the American military. With an eye toward history and international law, American military officers had accepted the fact that military governance was a necessary and inevitable task. They knew that, trained or not, military officers would be required to carry out the duties of occupation. Having clearly absorbed the lessons from Colonel Hunt's World War I generation, they preferred to avoid similar on-the-job frustrations. This generation of officers intended to be better prepared.

A Soldier's Job? The Civil-Military Debate

The rise of the military government schools was the high-water mark in American military learning for such duties. Yet no sooner had the Army decided to embrace and institutionalize the mission than they were stymied by a political and philosophical debate at the highest levels of government and American society. Should the American government be in the business of training "military governors"? The very term evoked powerful emotions in the American psyche. Certainly American officers had performed such missions in the past, but had they not done so reluctantly and only out of necessity? Developing a formal military government program and actually planning for the task seemed a step toward imperialism that many political leaders and members of the American intellectual elite were not prepared to take. Newspaper articles began to surface describing the Army's new "school for Gauleiters," and the debate intensified.⁶⁰

Among the president and other government leaders, the necessity for someone to perform such postwar functions was not in dispute; the question was, who should do it? For those who felt it should be a civilian function, the argument centered on fears of military dominance and an erosion of civilian control over military affairs. The president came down on the side of civilians yet directed the military to take control of training. Meanwhile, for the military, the question was one of competence and efficiency. If the military were to be required to govern, they wanted to be prepared; and ideally, they wanted to be in charge. That these officers had taken note of the frustrations of previous generations was clear in their arguments.

One memo written by the director of the Military Government Division in 1942 stated,

> If there is one outstanding lesson to be gained from prior American experiences in military government, it is the unwisdom of permitting any premature interference by civilian agencies with the Army's basic task of civil administration in occupied areas . . . In those important American experiences in military government—three in number—where civilian influence was permitted to be exercised, the results were, respectively, demoralizing, costly and ludicrous.⁶¹

Ironically, the author of the memo, Colonel Jesse Miller, was more civilian than military officer. He had extensive experience with the JAG during World War I and had also served as a civilian lawyer-adviser to the PMG on the subject of internment of aliens. He was given a commission as a colonel when hired by General Gullion to help with the curriculum at the Charlottesville school and so became intimately involved in the civil-military debate. The memo goes on to outline the frustrating experiences of Army officers in the Civil War, the Philippines, and World War I.

On September 4, 1942, in an attempt to clarify their intentions and curb the debate, PMG Gullion and his staff prepared a "Synopsis of Military Government" that detailed their philosophy for postwar occupation and outlined their activities to date. The synopsis claimed that "any oc-

---

60. Coles and Weinberg, *Civil Affairs*.
61. Memo, Colonel Jesse Miller for Colonel Edward S. Greenbaum, OUSW, July 23, 1942, PMGO files, 014.13, MG, as excerpted in Coles and Weinberg, *Civil Affairs*.

cupation of hostile or Axis-held territory may be divided into two phases: (a) a period of military necessity and (b) an ensuing period when military necessity will no longer exist."[62] In each phase, trained personnel would be required. The synopsis went on to outline the steps the War Department was taking to train officers for Phase I and to develop a "technique" for transferring control to civilian experts who would lead efforts in Phase II.

Having spelled out this division of labor, the synopsis called on "other agencies of the government" to submit a list of qualified personnel to participate in Phase II and to begin to study the problem of postwar occupation: "The War Department has heretofore suggested to the State Department and the Board of Economic Warfare certain studies in the fields of international law and economics. A need for research in other fields exists, and studies concerning them are to be requested."[63] Unfortunately, the "other agencies" of the government were far less organized or capable of the sort of operational mission planning and logistics that were to be required. No help came.

The Washington debate spilled over to the war front with somewhat disastrous consequences in November 1942. The Allied invasion of North Africa (Operation TORCH) would have benefited from the infusion of the many newly trained military governors—had they been allowed to deploy. Instead, 85 Charlottesville graduates remained stateside as the Allied invasion began. The president, having claimed that "the governing of occupied territories may be of many kinds but in most instances it is a civilian task,"[64] appointed Mr. Robert Murphy as the operating executive, head of the Civil Affairs Section, and advisor for civil affairs under General Eisenhower. Thus, instead of newly trained "military governors" taking the lead, a split authority developed between Murphy and Eisenhower. The result was a hodgepodge of civilian agencies, including the Board of Economic Warfare, the Lend-Lease Administration, and the State Department Office of Foreign Relief and Rehabilitation, all of whom attempted, somewhat autonomously and universally ineptly, to "manage" civil affairs in North Africa. Yet with over 30,000 tons of relief supplies arriving each month, only the Army had the capacity to meet the logistical requirements. Without a clear

---

62. The "Synopsis" of the War Department's Program for Military Government, September 4, 1942, PMGO files, 321, PMGO & MG, in Coles and Weinberg, *Civil Affairs*.
63. Ibid.
64. Quoted in Brigadier General Dennis A. Wilkie, "Deja Vu: 'Who's in Charge?'" *Scroll and Sword, Journal and Newsletter of the Civil Affairs Association* 56, no. 3 (2003).

mandate to take control over the operation, however, the Army was just another actor on the stage.[65]

Even as Operation TORCH continued, the Army began to lobby for more control over future operations. The secretary of war, along with General Gullion, General Marshal, and others, convened a series of conferences to discuss the best course of action in light of the North Africa experience. As a result, in January 1943, the War Department adopted a new policy to ensure that future war planning incorporated details on humanitarian relief for civilians to be carried out by military, not civilian, forces. Accordingly, military logistics planners were to include provisions for relief supplies such as food and medical equipment for civilians. To further institutionalize this policy, the War Department established the Civil Affairs Division (CAD) as part of the General Staff. With Major General Hilldring at the helm and with the requirement of operating side by side with the command center in the General Staff, the establishment of the CAD would ensure that civil affairs issues would be part of all subsequent campaign planning. More than just a Band-Aid fix, this move reflected the ability of the Army to learn in action, from both historical experience and from direct feedback from the field.

While the World War II Army was eager to take on this task in the name of efficiency, the sentiments of the civilian leadership that such duties were an inappropriate use of military forces became more accepted by military leadership over time. In the decades following World War II, civil affairs units were never fully institutionalized into the mainstream military. Warfighting doctrine conveniently ignored most problems associated with postwar issues, and the small civil affairs units that remained were relegated to the reserve component. Today, approximately 95 percent of civil affairs personnel are reservists. Unfortunately, a competent civilian counterpart schooled and capable of assuming this role also never emerged. The debate over civilian versus military control in such operations and the struggle to improve civilian capacity for stability and reconstruction continues to this day.[66]

---

65. Rick Atkinson, *An Army at Dawn: The War in Africa, 1942–1943*, Liberation Trilogy, vol. 1 (New York: Holt, 2007); Wilkie, "Deja Vu: 'Who's in Charge?'"; Ziemke, *The U.S. Army in the Occupation of Germany, 1944–1946*.

66. Michele Flournoy, *Interagency Strategy and Planning for Post-Conflict Reconstruction*, draft white paper produced for the Post-Conflict Reconstruction Project, March 27, 2002 (Washington, DC: Center for Strategic and International Studies); Clark Murdock, Michele Flournoy, and Chris Williams, *Beyond Goldwater-Nichols: Defense Reform for a New Strategic Era* (Washington, DC: Center for Strategic and International Studies, March 2004).

Institutional Learning and Political Resistance

The actions taken by the Army from the end of World War I through World War II do not reflect a military that was incapable of learning from its past or from the experience of others. The establishment of the education and training programs for military government and military police reflected a concerted effort to improve the Army's capacity for occupation duties, including governance and reconstruction. The Army was simply loath to have its officers perform a mission for which they were unprepared—again.

The institutionalization of past lessons was facilitated by structural changes in the military organization that were begun during the short American involvement in World War I and as directed in the National Defense Act of 1920. Whereas pleas from officers of previous generations to improve education for such duties had repeatedly gone unheeded, veterans from World War I were able to inject their ideas into the new planning processes, thus elevating the importance of the civil affairs/military government concept. The new focus on planning and the organizational and educational structures that were built around it improved the institutional capacity of the Army to process historical lessons learned and to generate and disseminate new ideas. In short, the Army had taken a few more steps toward becoming a "learning organization," and, as a result, civil affairs officers were born.

The military would continue to struggle against internal and external resistance to learn from its own past. While it was clearly making progress in developing institutional processes for capturing lessons learned, it was often hindered from translating insight from its more controversial roles, such as counterinsurgency and nation building, into doctrine, training, and education. From an organizational learning perspective, we see how institutional processes designed to capture experience can overcome internal cultural resistance to change but then still be stymied by outside or top-down political pressure. This politicization is especially acute in the delicate civil-military relationship in which civilian leaders must respond to public opinion about the "appropriate" use of force. The debates over military governance in World War II would sound all too familiar to those debating the use of military forces for stability and reconstruction missions today.

CHAPTER 3

Vietnam to Iraq: Debating the "New World Order"

The last two months of 1989 witnessed two significant events—one marking the end of an era, the other a beginning. On November 9, the world watched in stunned anticipation while East German citizens scaled the Berlin Wall and thus began to dismantle, both symbolically and physically, the structure that had held the Cold War order in place. Meanwhile, 5,000 miles away, the U.S. Army's 82nd Airborne was preparing for its first combat jump since World War II. Operation Just Cause, the one-week mission to oust Panamanian dictator Manuel Noriega, challenged the military capabilities of the world's remaining superpower and foreshadowed the series of post–Cold War contingencies the United States was about to face. Conducted with a joint U.S. force in an urban environment amid thousands of innocent civilians and multiple television crews, this operation was like nothing this Cold War military had seen or trained for.

The generation of officers who planned and led Just Cause, like those who led Desert Storm, had spent their careers preparing for major combat against Soviet forces. As this chapter shows, however, their subordinates would not have such clearly defined careers. Many of the young platoon leaders, company commanders, and battalion commanders who together led troops into the streets of Panama in 1989 would go on to serve in the streets of Somalia, Haiti, Bosnia, Afghanistan, and Iraq. Ironically, their operational experiences would seem strikingly disconnected from the strategic and political debates echoing in the halls of Washington and inside the Pentagon. While the military leadership, their civilian masters, and other pundits debated such issues as which military bases needed closing, how best to spend the "peace dividend," gays in the military, and whether or not women soldiers should train side by side with men, these officers were learning in the streets of Somalia, Haiti, and Bosnia just how complex this "new world order" was going to be.

This chapter sets the stage for understanding the post–Cold War political forces acting on the learning system. It provides a review of the operations conducted in this era and the accompanying strategic and political debates regarding the role of the military. From the perspective of unit commanders in the 1990s, the disconnect between politics and strategy, on the one hand, and operational reality, on the other, generated ambiguity over military roles and missions—an ambiguity that reverberated strategically and in the field. Political debates over "exit strategies," "mission creep," "casualty aversion," and "nation building" were informed by the devastating failures in both Vietnam and Somalia. The resulting "Vietmalia syndrome" translated to mixed messages regarding strategic and operational objectives—even as U.S. military units continued to deploy for MOOTW missions at a steady pace. Commanders faced with this strategic-level ambiguity often stretched the tensile strength of policies and regulations as they devised their own operational objectives and tactical rules of engagement to lead their units on ill-defined "other-than-war" missions.

Expectation and Reality: Peace Dividends versus the "New World Order"

Unrealized expectations about the need for U.S. troops resulted in a great deal of frustration throughout the 1990s among the military and civilian leadership. As the Berlin Wall fell, many civilian and military leaders assumed that the end of the Cold War meant an end to the need for a strong, globally deployed U.S. military. For many Americans, and especially for the American leadership, this "new world order" meant that thousands of troops stationed overseas as a counterweight to Soviet military strength should and could come home. Moreover, because the United States was released from the burden of an expensive arms race with its communist adversary, military spending could be drastically reduced to create a "peace dividend" for use on domestic issues. Accordingly, both presidential candidates in the 1992 election, George Bush and Bill Clinton, supported similar military reductions of approximately 25 percent.

In retrospect, this image of a peaceful, less-threatening global order may seem naive or premature, but the notion reflected a veritable consensus at the time among policymakers in Washington. Plans for downsizing and anticipation of peace dividends gained momentum as early as

1991, even as American military forces were fully engaged against Iraq in the Persian Gulf War.[1] Some members of Congress even called for military budget cuts as high as 50 percent.[2] "I suspect," said Colin Powell, "I must be the only Chairman of the Joint Chiefs of Staff in history who has ever testified before Congress on how to cut the force that was, at that moment, fighting a war 8,000 miles away."[3] Accordingly, and without much debate, the armed forces were swiftly cut by 25 percent, from approximately 2.1 million troops to 1.6 million, and the numbers continued to shrink in the subsequent years.

By 1997, with over 104,000 soldiers, sailors, airmen, and marines stationed in or deployed to foreign countries—including 9,000 troops in the Balkans and over 3,000 in the Persian Gulf—the U.S. military had still shrunk to approximately two-thirds the size of its Cold War peak.[4] Total troop strength was 1.4 million compared to the 2.2 million Cold War high, and the budget was approximately 38 percent smaller in real terms.[5] This relatively uncontroversial slashing of the Department of Defense (DoD) budget reflected the widely accepted expectation that the large Cold War military was no longer needed. The question was, what would take its place, and what would be its role?

As these events indicate, the initial post–Cold War debate focused more on quantitative changes rather than qualitative ones. As General John Shalikashvili explained, "the great issues of defense planning for an uncertain world have often been reduced to one simple question: How much is enough?"[6] Since most agreed that the military would have a dramatically decreased post–Cold War workload, the downsizing was fairly uncontroversial. The U.S. military has a long tradition of demobilization after war, so why should the end of the Cold War be different? However, although military leaders may have been prepared for a *diminished* role,

---

1. Colin Powell, "News Briefing: 1993 Report on the Roles, Missions and Functions of the Armed Forces, Submitted by Colin Powell, Chairman of the Joint Chiefs of Staff" (Washington, DC: Office of the Assistant Secretary of Defense [Public Affairs], February 12, 1993), http://www.fas.org/man/docs/cormg3/brief.htm.

2. Janine Davidson, "Doing More with Less? The Politics of Readiness and the U.S. Use of Force" (master's thesis, University of South Carolina, 2002).

3. Powell, "News Briefing: 1993 Report on the Roles, Missions and Functions of the Armed Forces."

4. Department of Defense, "Active Duty Military Personnel Strengths by Regional Area and by Country (309a)" (Washington, DC: Washington Headquarters Services, Directorate for Information Operations and Reports, September 30, 1997).

5. William S. Cohen, *Report of Quadrennial Defense Review,* May 1997; and *CDI Military Almanac* (Washington, DC: Center for Defense Information, 1999).

6. General John M. Shalikashvili, USA, Chairman, Joint Chiefs of Staff, address to the Council of Foreign Relations, New York, November 7, 1996.

it seems clear that they were not prepared for a *changed* one. The political pressures to send troops abroad during the 1990s meant that the U.S. military would need to adapt.

Qualitative Changes and Trends in Post–Cold War Military Operations

While the global "footprint" of U.S. forces diminished rapidly as Cold War troops redeployed from U.S. bases abroad, four significant qualitative shifts in military operations began to emerge. These shifts related to (1) the type and purpose of the mission, (2) the presence of nongovernmental organizations (NGOs) and other nonmilitary actors, (3) the types of troops sent, and (4) the average length of the engagement. Adjusting to these changes challenged military paradigms and contributed to a perception (and another debate) that the military was headed for a crisis in "readiness."[7]

The first qualitative change in military operations was in the nature of the missions being performed. Compared to the two decades prior, the U.S. military in the 1990s was involved in significantly more open-ended, low-level interventions and peacekeeping operations. During the full Cold War era, only three peacekeeping operations were executed. They were all in the Middle East region: two in Lebanon and one in the Sinai. In contrast, between 1989 and 2000, the United States participated in five peacekeeping operations, in four separate regions: Europe (the former Yugoslavia), Africa (Somalia and Angola), the Western Hemisphere (Peru and Ecuador), and Southeast Asia/the Pacific (East Timor). These peacekeeping missions and other post–Cold War operations were significantly different in character from operations in previous decades.[8]

Critical distinguishing characteristics of these post–Cold War operations were that they were usually conducted with multilateral forces and initiated for major humanitarian considerations. Less than 5 percent of military events during the Cold War were undertaken in coordination with another outside party or as part of a United Nations operation. In contrast, in the 1990s following the Gulf War, 35 percent of U.S. military operations were multilateral in nature. In addition, many of these deployments occurred in response to human suffering during or in the aftermath of violent conflict. Thus, in addition to keeping the peace, these

---

7. Davidson, "Doing More with Less?"
8. Ibid.

operations usually required military forces to assist refugees and internally displaced persons.

This change led, in turn, to the second major shift in the new era, interaction with NGOs and intergovernmental organizations.[9] The presence of refugees and other civilians in the "battlespace" and the accompanying transnational nature of many of these conflicts presented challenges to military operations. Civilians fleeing their homes often sought refuge in neighboring countries, where, unfortunately, they were not always welcome. These issues meant that the military were increasingly required to work with nonmilitary and allied actors, such as USAID, NGOs, and various NATO partners. That this is a fundamental shift in focus for the U.S. military is reflected in the observations made by a U.S. general in Operation Uphold Democracy in Haiti. He claimed that throughout his career, he had prepared himself for the time when he would command a major military operation. He had always envisioned landing at the site with the elite 10th Mountain Division on his left and the 101st Airborne on his right. When he arrived in Haiti, however, he found he had USAID on his right and a group of NGOs on his left.[10]

The third significant shift challenging the U.S. military was in the types of troops involved. Whereas most irregular events during the 1970s and 1980s were small-scale events often conducted by Special Forces or small groups of "advisors," the scope of peace operations in the 1990s required larger-scale deployments of troops from regular military units as well as the reserve component. Infantry and Armor units, not Special Forces alone, conducted the majority of the rotations to Somalia, Haiti, and the Balkans.[11] These back-to-back contingencies made peace operations the primary role for this generation. In 1994, for instance, over 40 percent of the troops sent to Haiti from the 10th Mountain Division had also served in Somalia.[12] This experience was a critical element in the professional development of a generation of U.S. military officers, espe-

---

9. Ibid.
10. Story related by Sherri W. Goodman, Deputy Undersecretary of Defense for Environmental Security, in a presentation given at the Women in International Security Summer Symposium, Pentagon, Arlington, VA, June 20, 2000.
11. The following units served in the Balkans: the 1st and the 3rd Infantry Divisions, the 1st Armored Division, the 1st Cavalry Division, the 10th Mountain Division, and the 101st Airborne Division.
12. Walter E. Kretchik, Robert F. Baumann, and John T. Fishel, *Invasion, Intervention, "Intervasion": A Concise History of the U.S. Army in Operation Uphold Democracy* (Fort Leavenworth, KS: U.S. Army Command and General Staff College Press, 1998), http://www.globalsecurity.org/military/library/report/1998/kretchik-chapter5.htm.

cially for the Army. It is this generation that comprised the leadership for the irregular challenges of the 21st century: Afghanistan and Iraq.

The fourth post–Cold War trend was the change in length of the average military engagement. Military operations carried out during the 1990s were significantly longer than similar types of events during the Cold War. None of the 12 MOOTW offensives during the Cold War (including strikes, invasions, and interventions) lasted longer than one year.[13] In fact, "strikes" were generally one-day affairs, and only five of the seven interventions lasted less than 90 days. The most major and more recent invasions of Grenada and Panama lasted less than a month each.

In contrast, prior to the invasion of Afghanistan, the average length of a post–Cold War military operation was 902 days (2.5 years), and the median was 730 days (2 years).[14] While the U.S. military intervention in Haiti lasted less than one year (the "police" aspect continued for much longer), operations begun in Bosnia in 1992 and in Kosovo in 1998 were still in progress as the bombs began to fall in Afghanistan in 2001 and in Iraq in 2002.[15]

This shift in mission duration came as a shock to the post-Vietnam generation, most of whom had spent the majority of their careers *training* for major theater war against a Soviet adversary, not actually *deploying* for real-world operations. One visitor to the Third Infantry Division in Bosnia observed, "Every other mission in most soldiers' memories had been a simulation or a game—even the Gulf War had turned out to be, as one of the veterans said to me, hardly more than a training exercise with consequences."[16] Thus, at the end of the Cold War, most troops were understandably accustomed to "real-world" military operations being executed swiftly, if at all. Restricted rules of engagement as a result of nonmilitary actors on the "battlefield" added additional frustration for military forces otherwise organized, trained, and equipped for full-scale "force-on-force" warfare.

Finally, contributing to the sense of frustration and overtasking during the 1990s was the fact that these long-term missions were accomplished

---

13. Cold War offensive operations included seven *Interventions:* Lebanon,1958; Congo, 1960; Dominican Republic ("Powerpack"), 1965; Congo, 1967; Egypt ("Nickel Grass"), 1973; Nicaragua, 1981; and the Philippines, 1989; three *Offensive Strikes:* Libya (Gulf of Sidra), 1981; Iraq ("Praying Mantis"), 1988; and Libya, 1989; and two *Invasions:* Grenada ("Urgent Fury"), 1983; and Panama ("Just Cause"), 1989. Davidson, "Doing More with Less?"

14. Ibid.

15. Lester H. Brune, *The United States and Post Cold War Interventions* (Claremont, CA: Regina Books, 1998).

16. William Langewiesche, "Peace Is Hell," *Atlantic Monthly* (October 2001).

through the use of rotating units and the reserve component (Guard and Reserve units). In contrast to the previous generation of military members who served with their families in comfortable U.S. military installations in Allied countries such as Germany and Japan, 1990s military personnel were sent on numerous three- to six-month tours away from their home bases (and their families) to temporary "tent cities" in the Balkans or the Saudi Arabian desert. Even before the 12- to 15-month deployments to Iraq began in 2003, the constant deployment and redeployment to remote locations had created personal and professional hardships that military members in an all-volunteer force seemed less willing to endure.[17] Meanwhile, at the political and strategic levels, these qualitative shifts in the use of military forces challenged military paradigms about the "proper" use of American force and strained civil-military relations. In Washington and the Pentagon, these new missions fueled the debate over the proper roles and missions for the U.S. military.

Cold War Attitudes and Lessons of Vietnam

In order to understand the nature of the changes required for the military to adapt to these new missions, we must have a clear understanding of the military's mind-set at the time. As previous chapters have demonstrated, despite the long tradition of conducting small wars and nation building, the U.S. military has historically and institutionally demonstrated a preference for "big war." Small wars advocates, both civilian and military, including unit commanders serving on the American frontier, the Caribbean, and Vietnam, frequently battled with military strategists and intellectuals who promoted what Robert Cassidy calls the "big-war paradigm." "The U.S. Army has embraced a big-war paradigm at least since World War I," he writes,

> and more probably since the influence of Emory Upton during the last quarter of the 19th century. This preference for the conventional paradigm became embedded in U.S. military culture over time and by the time of the Vietnam War, this preference shaped the U.S. Army so much that it was unable to adapt itself to counterinsurgency, instead preferring to apply a big-war paradigm when it was entirely inappropriate.[18]

---

17. For an excellent and detailed description of the elaborate planning and complex logistics involved in rotating large units in and out of the Balkans, see ibid.
18. Colonel Robert C. Cassidy, "Prophets or Praetorians? The Uptonian Paradox and the Powell Corollary," *Parameters* (Autumn 2003); Robert M. Cassidy, *Peacekeeping in the Abyss: British and American Peacekeeping Doctrine and Practice after the Cold War* (London: Praeger, 2004).

Although this perspective is challenged by some, it is gaining credibility in light of ongoing operations in Iraq and elsewhere. Regardless of the outcome of the next round of Vietnam debates, it is clear that the experience and the collective ways in which the institution made sense of it had profound implications for MOOTW operations in the 1990s and beyond.

Despite the fact that operations in Southeast Asia reflected many of the characteristics of the small wars of the past, the American big-war paradigm was sustained in the decades following the Vietnam War. In fact, in many ways, the Vietnam experience seemed to reinforce the sentiment that the U.S. military ought not to be involved in such missions at all. The debates continue to this day between those who attribute the failure to the Army's refusal to adopt an effective counterinsurgency strategy and those who blame the civilian leadership for preventing the Army from executing a proper full-scale conventional war. The later school of thought was promoted in an extremely popular book, *On Strategy: The Vietnam War in Context,* written by Colonel Harry Summers and sponsored by the Army War College.[19] Summers claimed that the Army had failed in Vietnam because it misconstrued the conflict as an insurgency and dallied ineffectively with a poor counterinsurgency strategy. Had the Army been better able to resist the influence of civilian leadership and instead applied its tried-and-true big-war strategies, the outcome would have been more favorable. The book was praised by Army leaders and was incorporated into educational curricula at all the major Army officer schools.

The Summers thesis stands in sharp contrast to a number of academic analyses and official military studies published throughout the 1980s.[20] These studies shared the conclusion that the American failure in Vietnam was due to the failure to adopt an appropriate counterinsurgency strategy and stick with it. As Andrew Krepinevich writes, "in the Army's thinking, there was scant difference between limited war and insurgency." Thus, he continues, "Army operations in South Vietnam were oriented overwhelmingly toward the Army Concept, with its bias toward

---

19. Harry H. Summers, Jr., *On Strategy: The Vietnam War in Context* (Carlisle Barracks, PA: U.S. Army Strategic Studies Institute, 1982).

20. BDM Corporation, *A Study of the Strategic Lessons Learned in Vietnam: Omnibus Executive Summary* (Washington, DC: BDM Corporation [sponsored by the U.S. Army War College], 1980); Andrew F. Krepinevich, Jr., *The Army in Vietnam* (Baltimore: Johns Hopkins University Press, 1986); Joint Low Intensity Conflict Project, *Joint Low Intensity Conflict Project Final Report, Executive Summary* (Ft. Monroe, VA: U.S. Army TRADOC, 1986).

mid-intensity conflict, big-unit operations, and minimization of U.S. casualties through heavy firepower."[21]

Krepinevich's book *The Army in Vietnam,* along with other Army-sponsored studies, also concluded that because of America's demonstrated aversion to this form of conflict, future enemies were more likely to adopt similarly "irregular" strategies.[22] Thus, the most likely form of conflict for the 1990s would be that for which the military was least prepared: low-intensity conflict. Ironically, despite the overwhelming consensus of these official and unofficial analyses, the U.S. military overwhelmingly favored the Summers thesis instead. This thesis effectively became the conventional wisdom for the post-Vietnam generation of military leaders.

This failure to adapt might also be examined from the perspective of learning theory. In this case, due to institutional inertia and the big-war military culture, the bottom-up organizational learning cycle was effectively blocked when it came to limited war. As Andrew Krepinevich explains, the organizational processes and policies of the Army in Vietnam stymied its ability to learn.

> The attrition strategy developed by MACV provided for the scaling of priorities, incentives, and rewards according to how well units operated according to traditional Army principles. The result was that any changes that might have come about through the service's experience in Vietnam were effectively short-circuited by Army goals and policies: tours were kept short, field commanders' missions were kept simple (kill VC [Vietcong]), firepower was readily available to accomplish the attrition strategy, and inflated body counts were acceptable, if not officially promoted. By the same token, these goals and policies frustrated attempts to conduct counterinsurgency operations.[23]

This tendency for the Army to revert to its "normal" processes was described in a 1972 study conducted for the RAND Corporation by Robert Komer. In *Bureaucracy Does Its Thing,* Komer points out that in Vietnam, even though many in the individual bureaucracies knew what needed to be done and even though there were high-level policies in place articulating the right counterinsurgency strategy, individual organizations tended to revert to the tasks they were designed to conduct rather than

---

21. Krepinevich, *The Army in Vietnam.*
22. BDM Corporation, *A Study of the Strategic Lessons Learned in Vietnam;* Joint Low Intensity Conflict Project, *Joint Low Intensity Conflict Project Final Report.*
23. Krepinevich, *The Army in Vietnam.*

adapting to the circumstances on the ground. This led to suboptimal outcomes. "We fought the enemy our way," he writes, "—at horrendous cost and with tragic side effects—because we lacked the incentive and much existing capability to do otherwise."[24]

John Nagl echoes these themes in his examination of the learning system of the U.S. Army in Vietnam. He highlights a cultural predilection for "big war," but more important, he claims that the Army as an institution was incapable of transferring new knowledge from the field to the rest of the organization. Compared to the British military that were able to adapt over time to the changing circumstances in Malaya, the U.S. Army was simply not a "learning organization." The Army's systems and culture made it like a huge ocean liner set to course. Large course corrections were not easy to make—especially by junior or midlevel officers learning from experience in the field.[25] In other eras, on-the-job initiative, informal networks, and generational learning often filled the gaps for leaders faced with familiar experiences, but in Vietnam, the rich experiences by junior and midlevel officers in Vietnam were not analyzed and disseminated as formal or informal doctrine either during or after the fact.[26]

This is not to say that contemporaneous learning did not occur at all. As Dennis Vetock explains, the combination of the old methods of field reporting with the post–World War II emphasis on new analytical methods such as probability theory and operations analysis generated a great deal of "lessons" from the field. The degree to which these lessons were analyzed and disseminated, however, is still in debate. While some felt the massive amount of information was so unmanageable as to be useless, others felt that the new mathematical and analytical tools enabled the management of this information at a sophisticated level of data processing that heretofore had not been possible. According to Vetock,

> The analytical lessons applied to the Vietnam operations became another form of contemporaneous lesson learning. The data that underwent analysis served as distilled combat experience. Collected from the battlefields and transformed quantitatively into charts and

---

24. Robert Komer, *Bureaucracy Does Its Thing* (Santa Monica, CA: RAND, 1972).

25. Richard Duncan Downie, *Learning from Conflict: The U.S. Military in Vietnam, El Salvador, and the Drug War* (Westport, CT: Praeger, 1998); John A. Nagl, *Counterinsurgency Lessons from Malaya and Vietnam: Learning to Eat Soup with a Knife* (Westport, CT: Praeger, 2002). Likewise, Downie demonstrates that the Army's learning cycle was stymied during and after Vietnam and thus the Army was institutionally unable to absorb and process the experiential lessons vis-à-vis counterinsurgency.

26. Downie, *Learning from Conflict*.

statistics, such data revealed, under analysis, the trends and patterns of operational experience. Conclusions thus drawn constituted lessons.[27]

Unfortunately, at the unit level, it is easy to see how other systemic pathologies undermined these potential gains. First, given the career structure for unit officers and the pressures to meet certain goals, strong incentives existed to stretch the truth in battle reports. Second, while "lessons" may have been processed from the vast amounts of "data" collected, the individual personnel rotation system (one-year rotations of individuals rather than of units) worked against effective contemporaneous dissemination. This dilemma is the source of the common saying among officers that the United States fought in Vietnam not an eight-year war but, rather, a one-year war eight times. Finally, while the quantification of data may have helped leaders in Washington assess certain measurable benchmarks, it is not clear how such statistics helped to improve tactics, techniques, and procedures at the unit level. Frustrations by junior officers in Vietnam laid the cultural and intellectual foundation for dramatic post-Vietnam changes.

It is difficult to overstate the effect the Vietnam experience had on the U.S. military. The collective lessons learned by this generation were that Vietnam was an anomaly—"an exotic interlude between the wars that really count."[28] The fact that this was the same assessment made in so many of the other irregular conflicts the United States had conducted in its 200-year history seemed to go unnoticed—again. The convenient conclusion was that the military did not need to change based on Vietnam; in fact, adapting for such an unlikely type of conflict would be folly. Focusing on irregular or small wars issues would only detract from the greater, more serious threat—the Soviet Union. Thus, by embracing the Summers thesis, the military's collective faith in "big war" was reaffirmed, allowing it to move forward with renewed focus on the wars it would prefer to fight. Robert Cassidy explains, "The Army's preference for large conventional wars had not been altered as a result of that paradigm's failure in Southeast Asia."[29] This big-war perspective sustained after Vietnam and was reflected in the popular Weinberger Doctrine, which provided a strategic guidepost for the new generation of military leaders.

---

27. Dennis J. Vetock, *Lessons Learned: A History of Us Army Lesson Learning* (Carlisle Barracks, PA: U.S. Army Military History Institute, 1988).
28. Brian Jenkins quoted in Cassidy, "Prophets or Praetorians?" 137.
29. Ibid.

Weinberger-Powell Doctrine

One of the most enduring outcomes of the Vietnam War was the so-called Weinberger-Powell Doctrine. More of a "policy" than a "doctrine," it clearly articulated for this generation of officers the way in which the military preferred to fight. The guidelines were first articulated by former secretary of defense Caspar Weinberger in a speech to the National Press Club on November 28, 1984. General Powell was the secretary's senior military advisor at the time of this critical speech. The secretary's speech resonated with military leaders of Powell's generation who had served frustrating tours of duty as junior officers in Vietnam and who felt that the military had been set up for failure in that conflict by what one Air Force general labeled "appalling rules of engagement" dictated from Washington.[30] To this generation of officers, not being allowed to fight to win their way ensured the incremental escalation of the conflict ("mission creep") and the subsequent loss of more American lives.[31] They felt that ultimately this led to a lack of political and public support at home and more ambiguity about military objectives.[32]

Accordingly, the Powell Doctrine sought to avoid such mistakes by dictating the following criteria for the use of military force: (1) force should only be used as a last resort, (2) force should be employed so as to "overwhelm" the enemy (i.e., not incrementally), (3) the objectives for the use of force should be well articulated prior to engagement, (4) the rules of engagement must be clearly defined, and (5) an "exit strategy" must be planned to avoid "mission creep" and excessive casualties. In short, the military was loath to be put in a situation where they could not fight, win, and come home rather swiftly and to great public approval. This articulation of when and how the U.S. military would fight amounted to what one Army strategist derisively called a "rewriting of the military's 'union contract' with the American people."[33]

The overwhelming and swift military victory against Saddam Hussein in

---

30. General Merrill McPeak quoted in David Halberstam, *War in a Time of Peace* (New York: Scribner, 2001), 40.

31. The thesis that conventional warfighting methods, unencumbered by civilian intervention, would have won the war in Vietnam is best articulated in the popular book by Harry H. Summers, Jr., *On Strategy: The Vietnam War in Context*. See also Harry H. Summers, Jr., "A War Is a War Is a War Is a War," in *Low Intensity Conflict: The Pattern of Warfare in the Modern World*, ed. Loren B. Thompson (Lexington, MA: Lexington Books, 1989).

32. Halberstam, *War in a Time of Peace*. See also James Kitfield, *Prodigal Soldiers: How the Generation of Officers Born of Vietnam Revolutionized the American Style of War* (New York: Brassey's, 1997).

33. Colonel Richard Lacquement, U.S. Army, personal correspondence with author, 2006, Pentagon, Washington, DC.

1991 proved to these military leaders (both civilian and military) that the Powell Doctrine worked. President Bush claimed at the time, "We've kicked that Vietnam syndrome for good."[34] In retrospect, this statement was obviously premature. Unfortunately, as troops faced their next irregular foe in the streets of Somalia, the other ghosts of Vietnam were still quite present.

Vietnam + Somalia = "Vietmalia Syndrome" and a New Policy for Peace Operations

Lessons from Vietnam loomed large as the U.S. military waded delicately into the violent streets of Mogadishu in late 1992. The tragedy suffered by troops on this mission and the collective political shock and outrage by the American public and national security leaders only added to the pathologies of the "Vietnam syndrome." The combined lessons from these two operations created additional resistance to operations other than major war, in what Richard Holbrooke described as a new "Vietmalia syndrome." This syndrome would present an updated political backdrop of resistance to subsequent MOOTW missions.

The Somalia operation began on December 9, 1992, when the First Marine Expeditionary Force landed on the beach in Mogadishu. Greeted with cheers and smiles from starving Somali people, their mission, as they understood it, was to provide a safe and secure environment for the delivery of food and relief supplies. Although aid workers rejoiced that the streets had been made safe so their work could begin, U.S. military commanders on the ground did not miss the fact that the famine had been caused by warring clans and a complete breakdown of governance. Operational commander Colonel Greg Newbold (later promoted to lieutenant general) told reporters on December 10 that although his marines had met little to no resistance upon landing, "I think they will test us later."[35] "Later" came on October 3, 1993.

After nearly 10 months of what seemed to be a successful humanitarian operation to feed starving Somalis and bring security to a war-torn country, Americans watched in horror as the body of an elite American Army ranger was dragged through the streets of Mogadishu.[36] Having

---

34. President Bush's speech given in 1991, quoted in Michael Hirsh, "America Adrift," book review of *War in a Time of Peace* by David Halberstam, *Foreign Policy* (November–December 2001): 160.

35. "Marines Bring Order to Somalian Capital," *St. Louis Dispatch*, December 10, 1992, A1.

36. For a detailed account of the incident leading to the deaths of 18 U.S. Rangers, see Mark Bowden, *Black Hawk Down: A Story of Modern War* (New York: Atlantic Monthly Press, 1999).

shifted the focus of the military mission from delivering food and relief (United Nations Operation in Somolia [UNOSOM] I) to confronting the underlying conflict plaguing Somalia (UNOSOM II), U.S. military forces found themselves in what Major General Steve Arnold claimed "may not be war, but sure as hell ain't peace."[37] The military operation to capture Somali warlord Mohammad Farrah Aideed turned sour when a Black Hawk helicopter was shot down in downtown Mogadishu. Efforts to rescue the downed airmen resulted in a protracted urban firefight that ended in the deaths of 18 American servicemen, 2 of whom were posthumously awarded the Medal of Honor.[38] This tragedy evoked powerful Vietnam-era emotions in military and civilian leaders, as well as in American society in general, and so became a political and doctrinal turning point in America's approach to peacekeeping and MOOTW.

As the grisly images were broadcast around the world, the effect in Washington was immediate. Congress was outraged, and President Clinton was shocked. Invoking the War Powers Act for only the second time since its inception in 1973, Congress called for the immediate withdrawal of all U.S. troops from Somalia. For peacekeeping advocates in the new administration, the failure in Somalia could not have come at a worse time. Having taken office less than nine months earlier, President Clinton and U.S. ambassador to the United Nations Madeline Albright had already embarked on an ambitious plan to improve the global capacity for peace operations—a plan that hinged on an invigorated U.S. commitment to peacekeeping and improved U.S.-UN relations. In the wake of the tragedy on October 3, their plans faced vigorous opposition and scrutiny.

Clinton's Plan for Peacekeeping

President Clinton's emphasis on peace operations as a fundamental role for the U.S. military was best reflected in a commencement speech he gave at West Point in May 1993.

> You will be called upon in many ways in this era: to keep the peace, to relieve suffering, to help teach officers from new democracies in the ways of a democratic army, and still to fulfill the fundamental mission

---

37. Steven L. Arnold, "Somalia—an Operation Other than War," *Military Review* (December 1993).

38. Shugart-Gordon, the mock village used to train Army units for irregular warfare at the Joint Readiness Training Center (JRTC) in Louisiana, was named in honor of Medal of Honor recipients Randall Shugart and Gary Gordon, both of whom were killed while saving the lives of the downed airmen. JRTC is discussed in detail in chapter 7.

which General MacArthur reminded us of, which is always to be ready to win our wars.³⁹

Earlier that year, hoping to institutionalize a greater U.S. role in UN peacekeeping, President Clinton had turned his attention to U.S.-UN relations. Almost immediately upon taking office in January, with troops already deployed in Somalia, President Clinton issued Presidential Review Directive 13 (PRD-13). As a first step in recognizing peace operations as a critical task, this order directed officials to review American contributions to UN peacekeeping and make policy recommendations for a potential presidential decision directive (PDD).⁴⁰

U.S. ambassador to the United Nations Madeline Albright, who was eager to repair U.S.-UN relations and improve what she saw as a frustrating, ad hoc approach to international intervention, was supportive of the PRD process. In a speech delivered to the Council on Foreign Relations in New York in June 1993, Albright expressed her frustrations at the "near total absence" of contingency planning by the United Nations for peacekeeping operations. This shortfall resulted in "hastily recruited, ill-equipped and often unprepared troops and civilian staff." Canadian-UN peacekeeper Major General Lewis MacKenzie described the dilemma more plainly: "Do not get into trouble as a commander in the field after 5 p.m. New York time, or Saturday and Sunday. There is no one to answer the phone."⁴¹ Such "ad hocery" was no longer unacceptable, asserted Albright. The ambassador claimed that "the time has come to commit the political, intellectual, and financial capital that U.N. peacekeeping and our security deserve."⁴² Thus, by improving U.S.-UN relations, Albright aimed to improve the United Nations' capacity and increase its role in responding to global conflicts.

Under the direction of the National Security Council and in coordination with various government offices, such as the State Department, the Office of the Secretary of Defense, and the Joint Staff, the proposed PDD-13 outlined a process by which the United States would assist the United Nations in improving its operational peacekeeping capabilities. Three features characterized the impact PDD-13 was to have on the U.S.

---

39. Speech transcript available online at http://www.findarticles.com/p/articles/mi_m2889/is_n22_v29/ai_14110648/print.
40. PRD-13 is often erroneously referred to as PDD-13. PRD-13 was a directive to conduct a review that resulted in a draft report containing recommendations for a PDD. PDD-13 was drafted but never signed.
41. R. Jeffrey Smith and Julia Preston, "United States Plans a Wider Role in U.N. Peace Keeping," *Washington Post,* June 18, 1993.
42. Ibid.

military. First, the United States would agree to participate in UN operations when it was considered in U.S. interests, not just when the United States could make a "substantial military contribution," as U.S. law at the time stipulated. This exemplified Albright's proposed strategy of "assertive multilateralism." Second, the United States would commit staff and resources, including 20 State Department and DoD personnel, to improve the operational capacity of the United Nations' peacekeeping headquarters. Finally and most controversial, the U.S. military would agree to allow its troops to operate under "operational control" of a UN commander. On this last point, the ongoing operation in Somalia was held up as a successful example of how such a command-and-control arrangement could work.[43] Of course, following the tragic events in Mogadishu, instead of providing a positive example for the administration's policy, Somalia provided a warning. PDD-13 was never signed, and interagency policy making focused on developing a new product to articulate President Clinton's peacekeeping policy, PDD-25.

PDD-25, "U.S. Policy on Reforming Multilateral Peace Operations," was signed on May 3, 1994.[44] The White House press release announcing the policy claimed that the document "represents the first, comprehensive framework for U.S. decision-making on issues of peacekeeping and peace enforcement suited to the realities of the post Cold War period."[45] Building on the questions raised under PRD-13 and the lessons learned from Somalia, PDD-25 focused more on the criteria for employing U.S. forces in UN peacekeeping missions rather than on improving UN peacekeeping capacities in general. Chairman of the Joint Chiefs of Staff Colin Powell, who resisted increasing the U.S. military commitment for peacekeeping missions, had heavily influenced negotiations for both PRD-13 and PDD-25.[46] Thus, the final product bore a remarkable resemblance to the Powell Doctrine.

In case there had been any doubt about where the chairman of the Joint Chiefs stood on the issue of MOOTW, Colin Powell published an

---

43. Susan Rice, former Director for International Organizations and Peacekeeping, National Security Council (1993–95), and former Assistant Secretary of State for African Affairs (1997–2001), personal interview by author, 2004, Brookings Institution, Washington, DC.

44. U.S. President William Jefferson Clinton, "Presidential Decision Directive-25: Clinton Administration Policy on Reforming Multilateral Peace Operation" (Washington, DC, 1994).

45. "Statement by the Press Secretary: President Clinton Signs New Peacekeeping Policy," White House, May 6, 1994.

46. Rice, interview.

article in the *New York Times* openly asserting that ill-defined MOOTW-type missions were not an appropriate use of U.S. military forces.[47] This open opposition by an extremely popular sitting chairman fueled the controversy over a rising civil-military "gap" in which the military leadership was thought to be openly and effectively thwarting efforts by the president (their commander in chief) to reorient military efforts toward MOOTW in general and peace operations in particular.[48] Indeed, overall civil military relations throughout the decade were strained by the military's opposition to Clinton's efforts to integrate homosexuals into the ranks, as well as the fact that he had openly resisted serving during the Vietnam War. In this tense climate, the Clinton administration had a difficult time promoting peace operations as a primary element of national security strategy.

As Lieutenant Colonel Robert Cassidy claims, "the Clinton doctrine on the use of force stemmed largely from the Weinberger-Powell doctrine, modified for peace operations in the aftermath of Somalia."[49] In fact, the effect the failure in Somalia had on the military cannot be overstated. The degree to which it resurrected the "ghosts" of Vietnam and crushed motivation for military participation in peace operations is evident in this statement by General Shalikashvili two and a half years after the incident.

> [In Somalia] our armed forces and our U.N. partners got caught up in mission creep. In the end, lives were lost and public support was forfeited. Many argue that the operation in Somalia provided us a very good lesson: Keep the military mission clear and "doable," and then leave when it is done. In any case, today, the specter of Somalia hangs over our involvement in peace operations.[50]

---

47. Colin Powell, "U.S. Forces: The Challenges Ahead," *Foreign Affairs* (Winter 2002–3); Colin Powell, "Why Generals Get Nervous," *New York Times*, October 8, 1992.

48. Colonel Charles J. Dunlop, USAF, "Welcome to the Junta: The Erosion of Civilian Control of the U.S. Military," *Wake Forest Law Review* 341 (1994); Peter Feaver, *Armed Servants* (Cambridge, MA: Harvard University Press, 2003); Peter Feaver and Richard H. Kohn, "The Gap: Soldiers and Their Mutual Misunderstanding," *National Interest* 61 (Fall 2000); Peter D. Feaver and Richard H. Kohn, eds., *Soldiers and Civilians: The Civil-Military Gap and American National Security* (Cambridge, MA: MIT Press, 2001); Richard H Kohn, "Out of Control: The Crisis in Civil-Military Relations," *National Interest* 35 (1994); Colin Powell et al., "An Exchange on Civil-Military Affairs," *National Interest* 36 (1994); Thomas E. Ricks, "The Widening Gap between the Military and Society," *Atlantic Monthly* (July 1997).

49. Cassidy, *Peacekeeping in the Abyss*.

50. General John M. Shalikashvili, USA, Chairman of the Joint Chiefs of Staff, "Speech Delivered to Care 50th Anniversary Symposium," May 10, 1996, Washington, DC, http://www.defenselink.mil/speeches/1996/s19960510-shali.html.

The combined influence of the Powell Doctrine, the Vietnam syndrome, and the collective trauma suffered in Somalia is evident in PDD-25. For example, Powell's emphasis on well-defined objectives was reflected in the stipulation that "the role of U.S. forces is tied to clear objectives." His "exit strategy" meant that "an endpoint for U.S. participation [must] be identified" and that "the operation's anticipated duration [must be] tied to clear objectives and realistic criteria for ending the operation." Powell's preference for "overwhelming force" in combat operations was clear in the insistence that "there exists a determination to commit sufficient forces to achieve clearly defined objectives."[51] For military officers, PDD-25 was considered a fair antidote to Clinton's apparent determination to involve the military in more "quagmires."

Meanwhile, the Vietnam generation's obsession with public opinion was addressed by PDD-25's requirement that "domestic and Congressional support exists or can be marshaled." Finally, the aversion to casualties as reinforced by Somalia was addressed by the stipulation that "both the unique and general risks to American personnel have been weighed and are considered acceptable."[52] Thus, the president's efforts to improve UN peacekeeping through increased U.S. assistance had effectively morphed into a set of rules designed to *limit* U.S. military involvement in operations other than war. Although PDD-25 reflected the administration's determination to take peace operations more seriously, interagency efforts to address both the lessons learned from the Somalia tragedy and the associated pathologies that had developed in the American psyche from Vietnam to Somalia meant that PDD-25 laid the foundation for (or simply documented) what Richard Holbrooke would label the "Vietmalia syndrome."

In practice, "Vietmalia" meant that every mission would be conducted in an effort to avoid the traumas suffered in these two previous events. As Colin Powell had written in his memoir, "we [the U.S. military] would not quietly acquiesce in half-hearted warfare for half-baked reasons that the American people could not understand or support."[53] For Powell's generation of military leaders, only "real" war, defined by the use of overwhelming force against a peer competitor, could qualify as not being "half-hearted" or "half-baked." Peace operations, nation build-

---

51. "Clinton Administration Policy on Reforming Multilateral Peace Operations (PDD 25), Executive Summary" (Bureau of International Organizational Affairs: U.S. Department of State, February 22, 1996); Clinton, "Presidential Decision Directive-25."

52. "Clinton Administration Policy on Reforming Multilateral Peace Operations (PDD 25), Executive Summary."

53. Colin Powell, *My American Journey* (New York: Random House, 1995).

ing, counterinsurgency, and every military operation "other than war" in between simply could not qualify and were therefore suspect. Never mind that many military analysts and most military officers felt that "big war" would be the least likely threat in the post–Cold War era.[54] MOOTW were an inappropriate use of the U.S. armed forces and were to be avoided—the U.S. military's mission was *to fight*.

The Real Mission, 2-MTW, and the Readiness Debate

The military's cultural penchant for "big war," reinforced by the Vietmalia syndrome, informed the decade's security strategy and the military's definition of "readiness." Being "ready" meant that the United States could respond with overwhelming force to one conflict without leaving itself vulnerable to attack elsewhere. Planners scanning the globe identified two potential regions where such capabilities existed to challenge U.S. might: the Middle East and the Korean Peninsula. While planners were cognizant of other unstable and dangerous areas of conflict, they felt that compared to the sizes of the militaries and the potential hostilities toward the United States in Korea and the Persian Gulf, other regions presented smaller, less significant threats. Thus, as the 1993 "Bottom-up Review" stated, "U.S. forces will be structured to achieve decisive victory in two nearly simultaneous major regional conflicts and to conduct combat operations characterized by rapid response and a high probability of success, while minimizing the risk of significant American casualties."[55] This became known as the "two major theater war" (2-MTW) strategy.

Throughout the 1990s, every important official strategic document issued by the White House or the Pentagon clearly reinforced the focus on 2-MTW. Even PDD-25, the era's first document on peace operations policy, made a point to highlight the 2-MTW strategy. "As specified in the 'Bottom-up Review,'" it stated, "the primary mission of the U.S. Armed Forces remains to be prepared to fight and win two simultaneous regional conflicts."[56] Likewise, although President Clinton's National Security Strategy (NSS) consistently promoted peace operations as "an important component of our strategy," each version of the administration's NSS was also careful to point out that "the primary mission of our Armed

---

54. Deborah Avant and James Lebovic, "U.S. Military Attitudes towards Post-Cold War Missions," *Armed Forces and Society* 27, no. 1 (2000).
55. Secretary of Defense Les Aspin, "Bottom-up Review" (Washington, DC, October 1993).
56. Ibid.

Forces is not peace operations; it is to deter and, if necessary, to fight and win conflicts in which our most important interests are threatened."[57] Thus the message for military planners became a bit muddled. Peacekeeping was important but not a priority. The resulting Quadrennial Defense Review (QDR) and National Military Strategy (NMS) documents clearly reflected the DoD's understanding that peace operations were something the military would do if so ordered but not something for which they must significantly change their force posture or strategic planning.[58]

This dichotomy was reinforced in the independent Commission on Roles and Missions (CORM), established by the National Defense Authorization Act for Fiscal Year 1994. The focus of the CORM was on evaluating the joint interoperability of the services since the 1986 Goldwater-Nichols Act. It was to review the allocation of roles among the services and make recommendations for changes to current structure and processes that would enhance effectiveness. The CORM was significant for its emphasis on improving service responsiveness to the regional commanders in chief (CINCs) during real-world operations. Many of its recommendations were written into law (including the requirement to conduct the QDR). With respect to peace operations and MOOTW, however, the CORM report stated the following:

> Peace Operations. Currently, DoD regards peace operations as a subset of the broad category of operations other than war. However, peace operations hold the prospect for preventing, containing or ending conflict. They have the potential to preclude larger, more costly U.S. involvement in regional conflicts. We recommend differentiating peace operations to give them greater prominence in contingency planning.
>
> OOTW. In addition, we must be prepared to engage in the wide range of remaining OOTW tasks, such as humanitarian assistance

---

57. William Jefferson Clinton, *A Strategy of Engagement and Enlargement* (Washington, DC: White House, February 1996).

58. White House policy is formally articulated in the annual National Security Strategy (NSS), which in turn (in theory) informs the Quadrennial Defense Review (QDR) published by the Office of the Secretary of Defense (OSD) and the National Military Strategy (NMS) published by the Joint Chiefs of Staff. The requirement to publish an annual NSS was written into law by the 1986 Goldwater-Nichols Act, while the QDR requirement was written into law by the Military Force Structure Review Act of 1996 following the recommendations of the 1995 Commission on Roles and Missions. The Bush administration's "Base Force" report and the Clinton administration's "Bottom-up Review," both voluntarily accomplished by OSD during these two administrations, served as a model for the QDR concept.

and disaster relief. For these, we recommend limiting the use of military forces to military tasks where practical; broadening non-DoD capabilities for some OOTW functions; and improving interagency coordination. We must also ensure rapid reimbursement of DoD for unplanned peace operations and OOTW to prevent readiness problems among forces not engaged.[59]

Thus the CORM, like PDD-25, noted the importance of peace operations to the greater national security in general. By differentiating them as a special subcategory of MOOTW (a.k.a. OOTW), the CORM further highlighted the important role the military had to play in conducting them. Yet, although the military was encouraged to "give [peace operations] greater prominence in contingency planning," no specific changes in force structure or doctrine were recommended. Moreover, the "remaining" MOOTW missions were clearly considered the primary responsibility—both financially and in practice—of "other" agencies. The military would play a limited role, but the U.S. government was expected improve the capacity of other agencies to take over such tasks. The subsequent failure to enhance these other agencies while at the same time assigning them primary responsibility for these critical tasks would have devastating consequences in Iraq and Afghanistan. As discussed in chapter 6, military planners believed that other experts in civilian agencies would take care of the postconflict phase of the invasion. Of course, practically no one showed up from these agencies because these experts simply did not exist—at least not with the capability and capacity the military assumed.

Despite the lukewarm emphasis on peace operations suggested in the somewhat schizophrenic CORM (or perhaps as a result of the CORM's lukewarm, less-than-directive language), the 1997 Quadrennial Defense Review and the 1997 National Military Strategy downplayed peace operations and reaffirmed the emphasis on 2-MTW. The QDR stated clearly, "U.S. forces must be capable of fighting and winning two major theater wars nearly simultaneously."[60] The Joint Chief of Staff's NMS followed suit.

The military has an important role in engagement—helping to shape the international environment in appropriate ways to bring about a more peaceful and stable world. The purpose of our Armed Forces,

---

59. "Commission on Roles and Missions of the Armed Forces, Report to Congress, the Secretary of Defense, and the Chairman of the Joint Chiefs of Staff" (Washington, DC, May 24, 1995).

60. QDR, 1997.

however, is to deter and defeat threats of organized violence to our country and its interests. While fighting and winning two nearly simultaneous wars remains the foremost task, we must also respond to a wide variety of other potential crises. As we take on these diverse missions, it is important to emphasize the Armed Forces' core competence: *we fight*. That must be the primary consideration in the development and employment of forces.[61]

In light of the CORM and the well-articulated 2-MTW emphasis in national security and military strategy, the question turned to exactly how the military was to prepare for peace operations while also staying prepared for major theater war. On one end of the spectrum were those who called for changes in force structure, perhaps even specialized units or constabularies to carry out peace operations. On the other end were those who still warned that soldiers ought not to conduct such missions at all. As John Hillen warned, "soft" missions like peacekeeping would degrade readiness and warfighting skills: "When the time comes for these soldiers to fight again, the United States will learn a painful lesson: Soldiers make good social workers, but social workers make lousy soldiers."[62] On this point, many military officers who had actually conducted peace operations disagreed.

Major General Bill Nash, commander of the First Armored Division in the Balkans immediately following the Dayton Peace Accords, claimed that his soldiers actually benefited from their deployment: "We found that the improvements gained in difficult battle staff and command skills while in Bosnia far outweighed any losses in specific war-fighting tasks while deployed for peacekeeping . . . There is no doubt in my mind that we were a much more capable, fighting division after Bosnia than we were when we went there."[63] For Nash, education of commanders is more important than training lower-level troops.[64] General Nash would agree with Major General S. L. Arnold, commander of the 10th Mountain Division force in Somalia, that "well-trained, combat-ready, disci-

---

61. John M. Shalikashvili, "National Military Strategy: Shape, Respond, Prepare Now: A Military Strategy for a New Era" (Washington, DC: Department of Defense, 1997).

62. John Hillen, "Playing Politics with the Military," *Wall Street Journal*, December 5, 1996. Ironically, as chapter 6 demonstrates, many of the skills these soldiers learned preparing for and executing the long series of peace operations in the 1990s served them well in Afghanistan and Iraq.

63. Major General Bill Nash and John Hillen, "Can Soldiers Be Peacekeepers and Warriors?" *NATO Review* 49, no. 2 (2001), online edition.

64. General (Ret.) William Nash, USA, personal interview by the author, October 2003, Council on Foreign Relations, Washington, DC.

plined soldiers can adapt easily to peacekeeping or peace enforcement missions."[65] Thus, no major changes to doctrine or training should be required.

Such opinions, combined with the strategic focus on 2-MTW, further reinforced the big-war mind-set. If soldiers could so easily adapt during peace operations and then spring back into warfighting shape upon redeployment, then little substantive institutional change would be needed to accommodate these roles. Another report on roles and missions commissioned separately by the U.S. Army also articulated this growing conventional wisdom by concluding, "Forces should not be earmarked for peace operations nor should new forces be created."[66] In the end, a de facto middle ground prevailed. No substantive changes were made to force structure that could be said to specifically target the requirements of MOOTW, but neither were peace operations abandoned as a role for U.S. military troops.[67] By 1996, over 20,000 American military members were conducting MOOTW around the world.[68]

As the tension between the 2-MTW strategy and the operational realities continued, members of Congress began to question the "readiness" of the U.S. military to conduct major theater war. Between 1995 and 1999, a series of reports from the General Accounting Office reflected the overall politicization of this issue.[69] Each of these reports had a slightly different research question, but all were designed to determine the degree to which these nonstandard missions were affecting the abil-

---

65. Quoted in Wray Johnson, *Vietnam and American Doctrine for Small Wars* (Bangkok: White Lotus Press, 2001).

66. Vector Research, "The 21st Century Army: Roles, Missions and Functions in an Age of Information and Uncertainty" (Ann Arbor, MI, 1995).

67. For an in-depth look at the struggle to change Army structure in the post–Cold War era, see Tammy Schultz, "Ten Years Each Week: The Debate over American Force Structure for Winning the Peace?" (PhD diss., Georgetown University, 2005); and the Pulitzer Prize winning series Price of Power, including Thomas E. Ricks, "The Price of Power—Ground Zero: Military Must Change for the 21st Century—the Question Is How," *Wall Street Journal*, November 12, 1999; Thomas E. Ricks and Anne Marie Squeo, "The Price of Power—Sticking to Its Guns: Why the Pentagon Is Often Slow to Pursue Promising Weapons," *Wall Street Journal*, October 12, 1999; Carla Anne Robbins, "The Price of Power—Ultimate Threat: U.S. Nuclear Arsenal Is Poised for War—Is It the Right One?" *Wall Street Journal*, October 15, 1999.

68. Department of Defense, "Active Duty Military Personnel Strengths by Regional Area and by Country (309a)."

69. Congressional Budget Office, "Making Peace while Staying Ready for War: The Challenge of U.S. Military Participation in Peace Operations" (Washington, DC, 1999); General Accounting Office, "Military Operations: Impact of Operations Other than War on the Services Varies (GAO/NSIAD 99-69)" (Washington, DC, May 1999); General Accounting Office, "Peace Operations: Effect of Training, Equipment, and Other Factors on Unit Capability (GAO/NSIAD-96-14)" (Washington, DC, October 1996).

ity of the U.S. military to maintain readiness for its "real" mission of major theater war. Taken together, the results were mixed. Although findings indicated wear and tear on equipment, others revealed that commanders felt the experience had been good exposure for their troops. The net result was a sense that regular military units could perform these missions and still maintain readiness, given enough time beforehand to train for tasks specific to peace operations and time again upon redeployment to retrain for combat duties.

In sum, the high-level debate over roles and missions and the ambiguity concerning the priority that should or should not be given to peace operations were informed by the Vietmalia syndrome and subject to politicization. The message was that the U.S. military was designed for the role of fighting and winning the nation's (big) wars. Powell's generation felt that strong leadership and a focus on big war had led the military to a long, slow recovery from the humiliating defeat suffered in Vietnam. They would not repeat the same mistakes. The result was an understanding that peace operations would be conducted if so ordered, but not at the expense of "readiness" for the 2-MTW role. Moreover, when these missions were conducted, the military would clearly prescribe objectives and military tasks and would avoid casualties, so as to ensure continued domestic public support. These messages resonated clearly with commanders in the field during operations throughout the 1990s and continued to define the organizational culture of the U.S. military as President Bush entered office in 2000. Such paradigms would not serve America's troops well as they deployed to Afghanistan and Iraq and became embroiled in the largest stabilization and reconstruction operations in the history of the all-volunteer force.

In the Trenches: Vietmalia in Action from Haiti to Iraq

Commanders in the field from Haiti to the Balkans were well aware that "Vietmalia" meant a collective aversion to both nation building and casualties. This translated to limited mandates for military operations and to political pressure on commanders to implement stern measures for force protection. Unfortunately, stabilizing a country is much more difficult when troops are prohibited from interacting with the population, addressing grievances, or helping restart the economy. In the peacekeeping scenarios of the 1990s, this meant that the strategic objectives of the mission often conflicted with the political requirements to avoid risky scenarios or any specific duties, such as governance or reconstruction, which might resemble "nation building." For some troops and their commanders, this was a great source of confusion and frustration.

Lieutenant General Greg Newbold was the Marine Expeditionary Unit commander for the mission to Somalia in 1992. As he recounted in a series of interviews in 2005, even the limited humanitarian mission they were assigned was stymied by inconsistencies in policies, resources, and directives.[70] His official orders were so nebulous that he had to call one of his stateside colleagues and ask him to watch President Bush's televised address to the nation and relay him the details so that he could infer from the president's remarks what his unit's operational objectives might be. Once on the ground, the marines' ability to maintain law and order or to get at the root causes of the conflict was stymied by a lack of resources, relevant doctrine, and political-cultural understanding of the environment. Moreover, the marines had no legal authority to arrest and detain criminals and warlords. Even if they had arrested them, there were no jails in which to lock them up and no indigenous government to which they could be delivered.

Similar issues plagued U.S. troops in Haiti and the Balkans. One after-action report from the Haiti mission observed, "Although the strategic objective of restoring democracy did not change, nor the operational objective of establishing a secure and stable environment, the supporting objectives to both became fuzzy." The result was that commanders took disparate approaches in translating the broad strategic objectives (i.e., to "restore and uphold democracy") to operational orders and specific rules of engagement. Rules of engagement with respect to the use of force varied from unit to unit. Specifically, whereas the marines were instructed to use deadly force in self-defense, troops in the Army's 10th Mountain Division were not granted the same authorization.[71]

Further limiting the actions of military forces in Haiti was the awareness by commanders that incurring American casualties as happened in Somalia could result in diminished public support for the mission and premature redeployment of forces. This fear of casualties translated to prohibitions on disarming local militants and to the limitation of troop patrolling to daytime patrols only. Meanwhile, the American political distaste for nation building prohibited the military from assisting in recon-

---

70. Lieutenant General (Ret.) Gregory Newbold, USMC, numerous personal interviews by author, June–July 2005, Arlington, VA. Newbold was a colonel at the time of the Somalia operation, in command of the Marine Expeditionary Unit that landed there.
71. Ibid.

struction operations or implementing rule-of-law measures. Many officers became frustrated by such limited mission objectives and proscribed military tasks, which they felt simply ensured that lawlessness would continue. This frustration often resulted in savvy attempts to circumvent or otherwise disobey the rules.

One of the most notorious cases of an officer taking issues into his own hands involved a junior officer stationed in Haiti with the 10th Mountain Division. Captain Lawrence Rockwood was court-martialed for disobeying orders and investigating human rights abuses at a local Haitian prison. Claiming that President Clinton had declared that the U.S. mission objective in Haiti was "to stop brutal atrocities," he did not understand why nothing was being done about the prison. He felt that his commanders were not carrying out the commander in chief's (President Clinton's) "mission intent." Moreover, he claimed it was his obligation under international law to investigate allegations of human rights abuses. Despite the fact that Captain Rockwood was given a human rights award from the New York Civil Liberties Union for his actions, the junior officer was court-martialed, found guilty, and discharged from the Army.[72]

The American distaste for nation building and the aversion to casualties evident in both Somalia and Haiti also frustrated commanders and hampered objectives in the Balkans. Political pressure to limit the nature and level of U.S. troop involvement began with negotiations at the Dayton Peace Accords. As Ambassador Richard Holbrooke recalls, the Americans at Dayton were plagued by memories of Mogadishu, "which hung over our deliberations like a dark cloud; and Vietnam, which lay further back, in the inner recesses of our minds."[73]

Pressure to limit military objectives also came from Congress. As General Bill Nash, the first commander charged with enforcing the Dayton Accords, prepared to lead his First Armored Division into Bosnia, he was subjected to a barrage of phone calls from Washington lawmakers and journalists asking him to clarify whether or not his troops would be carrying out a "nation-building" operation.[74] Fully aware of the precarious political perch on which he had been placed, General Nash assured anxious lawmakers that his goals were purely military. If news reporters filmed his troops "constructing a bridge here or there," the American

---

72. Ian Katz, "Depressed or Just Decent?" *Guardian*, May 30, 1995.
73. Richard Holbrooke, *To End a War* (New York: Random House, 1999).
74. Nash, interview.

people could be sure it was for tactical military purposes. Still, a perusal of the general's written "commander's intent" makes it clear that this experienced combat engineer understood the real links between reconstruction, economic development, and stability in a peacekeeping operation.[75] In his written guidance to his subordinate commanders, the general states, "Our implied task for the conduct of all operations is to facilitate non-military efforts toward infrastructure development, economic growth, and democratic practices."[76] That he also understood the political pressures vis-à-vis casualties is reflected in the last paragraph: "Ultimately, we will be judged by our ability to protect our forces in this highly uncertain, difficult and lethal environment. Therefore, I expect commanders and leaders to take the required measures to care for, secure, and protect their soldiers and organization."[77]

From Haiti to the Balkans, the American aversion to casualties became institutionalized by the U.S. military into rules of engagement that emphasized force protection. Whereas British and other NATO troops could be seen in soft caps sipping coffee with locals at neighborhood cafes, Americans were required to maintain full combat gear, including helmets, and spent most of their time in hardened garrisons having little interaction with the local population. This was especially true for rotations to Bosnia, in which the military had enough time to construct garrisons such as Eagle Base, about which one reporter wrote the following:

> The soldiers live in surprising isolation [on Eagle Base]. Those who go out on patrol are generally forbidden to sit in cafés, to shop in the local markets, or to socialize with the Bosnians. Two thirds of the soldiers never even go out on patrols, and live almost entirely restricted to the base, where they serve in "support" roles as guards and office workers, and seem to spend most of their leisure time eating and then trying to lose weight. Bosnia for them is a wooded campus that takes about twenty minutes to walk across. Although the environs of Eagle Base are safe, the installation is surrounded by an illuminated and guarded perimeter fence, seven miles around, with only three gates, all of which are strictly controlled.[78]

This emphasis on force protection meant that troops left garrison for military patrolling only. The military's role was to provide a safe and se-

---

75. Ibid.
76. General William Nash, *FRAGO 136, Commander's Intent* (January 1996).
77. Ibid.
78. Langewiesche, "Peace Is Hell."

cure environment, not to become cops or social workers. Thus, they resisted actively pursuing and arresting Bosnian war criminals, claiming that such duties were for police, not military forces. After much debate, they agreed to arrest such criminals if they were caught in the course of normal patrolling, but the military would not integrate the task into their mission orders. Likewise, reconstruction tasks were assumed to be the responsibility of civilian agencies and NGOs.

The debate over using soldiers as "social workers" did not abate. Still, a generation of troops learned how to support such missions, and as the following chapters reveal, the skills troops gained during their peacekeeping deployments in the Balkans became the basis for a new batch of doctrinal manuals and training scenarios. Most important, this experience was the experiential "handrail" on which they relied to guide their actions in postwar Iraq.

The Handrail

Veterans of the Balkans, such as major generals David Petraeus and Peter Chiarelli, were able to leverage their experience during the confusing post-combat phase of Operation Iraqi Freedom (OIF). Petraeus, who led the 101st Airborne Division during the initial invasion in 2003, found himself in charge of a large swath of northern Iraq after the fall of Saddam's army. The area included pro-American Kurdish regions as well as the large majority-Sunni city of Mosul, where Saddam's sons were killed by U.S. forces. Unlike other commanders who were shocked to learn that the mission did not end upon the fall of the enemy's military, General Petraeus arrived with a robust stability and reconstruction plan. Within two days of his troops landing in Mosul, Petreaus, who had served in the Balkans and who had written a PhD dissertation on counterinsurgency in Vietnam, was meeting with local leaders and plotting the stability and reconstruction for the region.

Likewise, Major General Peter Chiarelli, commander of the First Cavalry Division in Baghdad, took initiative to conduct an ambitious reconstruction program in his area of responsibility. For both commanders, "stability" was integrally connected to reconstruction and security. Troops rebuilt schools, hospitals, businesses, roads, and houses. They helped with the fall harvest, dug wells, provided jobs, and ran town hall meetings. They empowered brigade commanders to reconstruct the areas for which each was responsible and created command climates to show them the way. Petraeus's weekly PowerPoint briefing reminded his subordinate commanders, "We are in a race to win over the people. What have you and your element done today?"[79]

Similarly, Chiarelli emphasized six "lines of operation" as part of his "full-spectrum information operations" approach: combat operations, training and employing Iraqi security forces, essential services, governance, and economic pluralism. Claiming that "when we started building, they stopped shooting," Chiarelli spent the largest percentage of his time on reconstruction tasks, which he focused in the most run-down, insurgent-infested neighborhoods. The approaches of these two commanders were recorded in articles and speeches and widely disseminated. As discussed in the following chapters, these lessons ultimately became part of the new crop of doctrine, education, and training developed to address stability and counterinsurgency.

Summary

This chapter demonstrates the perennial disconnect between the political and operational levels in military affairs. Even as the president attempted to reorient national security policy toward peace operations, resistance to such missions was clear in both the civilian and military senior leadership. Still, a steady stream of operational missions exposed a generation of officers and their troops to the complexities of modern conflict. From Somalia and Haiti to Bosnia and Kosovo, the generation of leaders currently serving in Iraq and Afghanistan today "cut their teeth" on a decade of "other-than-war" missions. Lacking appropriate guidance, these junior and field-grade officers often made up their own rules and learned on the job. The question is, how has this experience transferred to the greater military organization?

The next two chapters address this question by examining the substantive changes in doctrine, education, and training since Vietnam and especially since 1989. Particular attention is given to the processes by which experiential learning was disseminated and the degree to which this dissemination was through formal or informal systems. The study reveals that despite the strategic and political ambiguity described in this chapter, this generation was able to produce a steady stream of updated doctrinal manuals to reflect their experiences. Moreover, the training

---

79. Numerous interviews with troops and officers of the 1st Brigade, 101st Airborne Division, site visit April 2004. Commander Major General David Petreaus, 101st Airborne Division, U.S. Army, lecture on reconstruction activities in northern Iraq, March 23, 2004, Washington, DC.

system put in place following the Vietnam War facilitated the dissemination of a vast amount of tactical-level lessons from this decade of operational experience. The remaining chapters describe the development of the modern military's learning system and analyze the system's capacity for organizational learning in today's complex stability and reconstruction operations.

CHAPTER 4

Learning to Learn: The Training Revolution in the Post-Vietnam Military

Previous chapters have revealed the difficulty the U.S. military has had in translating operational experience into concrete institutional knowledge. However, the military's experience since Vietnam suggests that this trend may be changing. In today's military—especially the U.S. Army—formal mechanisms and processes such as the Center for Army Lessons Learned, the process for after-action review, the Combat Training Center Program, websites, and military journals reflect a dynamic institution-wide learning system that enables the organization to leverage bottom-up experiential learning in ways unseen in previous eras. Today's Army is simply not the same one that failed to learn contemporaneously in Vietnam or the one that failed to capture useful doctrine from its decades on the frontier or in the Philippines. Although there is still much room for improvement, the military's processes and capabilities, which have been fundamentally and consciously transformed since the 1970s, provide an example for how a large, tradition-bound organization can promote contemporaneous, bottom-up learning and, in so doing, change the culture of the organization.

This chapter describes the various structural changes that were consciously put in place in the aftermath of the Vietnam War. These changes would have a mixed effect on the Army's ability to carry out post–Cold War MOOTW tasks but would continue to evolve and influence the organizational culture of succeeding generations. On the one hand, changes such as the creation of the all-volunteer force and the National Training Center would serve to reinforce the cultural paradigm that the military exists only to fight "big wars." This paradigm was also reflected in the resistance to change educational curricula at the professional service schools such as the Command and General Staff College and the war colleges, as discussed in chapter 5. On the other hand, changes such as the creation of the Light Infantry, the resurrection of the Special Operations Forces, and, most important, the learning-through-doing processes honed at the combat training centers would enhance the capacity of a generation of Army officers to learn from experience in ways that have gone relatively unrecognized. Over time, these changes would have a considerable effect on the organizational culture of the military, which would in turn become a key factor in turning the tide in Iraq. This shift in organizational culture, as evidenced by the active institutional promotion of what learning theorists call "communities of practice," the open and voluminous intellectual debates conducted across the institution, and the increasing and persistent refrain to "adapt" in the field, reveals a concerted effort by military leaders to "crack the code" for enabling a large, tradition-bound and complex organization to become a "learning institution."

Let the Healing Begin: Rebuilding the "Hollow Force" and the Evolution of the Modern Learning System

In the immediate aftermath of the Vietnam War and through the rest of the 1970s, the U.S. military was considered by many to have been "broken." Having relied on the draft and on the individual replacement policies throughout the highly unpopular war, the U.S. military ranks suffered an exodus of talent, especially at the critical NCO level. With nearly 50 percent of young military recruits thought to be drug users and with racial tensions running high, discipline was a full-time job for many officers.[1] These conditions, combined with aging equipment and a general lack of doctrinal focus, led Army chief of staff General "Shy" Meyer to report to President Carter (and later to Congress) that the military had become a "Hollow Force."[2]

The military officers who continued to serve after Vietnam were determined to rebuild and repair this force. They ended the draft, insti-

---

1. Colonel (Ret.) Robert Killebrew, USA, personal interview by author, June 10, 2005, Basin Harbor, VT; James Kitfield, *Prodigal Soldiers: How the Generation of Officers Born of Vietnam Revolutionized the American Style of War* (New York: Brassey's, 1997); Colonel (Ret.) Thomas Lency, USA, personal interview by author, 2009, U.N. Foundation, Washington, DC.
2. General (Ret.) Robert Scales, USA, *Certain Victory: The U.S. Army in the Gulf War* (Washington, DC: Potomac Books, 1998).

tuted a zero-tolerance policy on drug use, and created the professional "all-volunteer military." They developed new doctrine for high-tech warfare and designed training centers to test and institutionalize the new themes. Although these changes were all focused on rebuilding and refocusing the military on traditional warfare and decidedly away from counterinsurgency or what was termed "low-intensity conflict," the processes and structures developed would facilitate the rapid adaptation required in the early 21st century for COIN and stability operations.

The Development and Characteristics of the Combat Training Centers

The development of the combat training centers (CTCs) was a concerted effort to heal wounds from the Vietnam War. The leaders that developed the CTC program were intent on putting Vietnam as far behind them as possible. Military leaders such as General "Shy" Meyer, Major General Max Thurman, and General Paul Gorman were focused on rebuilding from the "Hollow Force" of the 1970s and preparing for war against the Soviet Army, the most serious threat the nation faced. Given this refocused priority on "big war," the Army built the enormous National Training Center (NTC) in Fort Irwin, California. Here, large tank formations could fight mock battles against a mock Soviet foe, and the Army could test its latest high-tech weaponry, instrumentation, and communications equipment.

What the CTC system added to the traditional system was a way to standardize and evaluate how units across the Army were trained for the common essential tasks ("knows" and "METLs" from the mission essential task list) as well as a means to generate "muscle memory" by "learning through doing" on a major scale. The success of this program would motivate the nonmechanized parts of the Army, including the Light Infantry, Air Assault, and Special Forces, to consider a version of the NTC designed to prepare soldiers for low-intensity conflict: the Joint Readiness Training Center (JRTC), discussed below. By the mid-1990s, when the operational tempo had increased, the CTC system became a place to rehearse prior to deployment and to filter new ideas from the field back to the rest of the Army (or at least to the next groups training at the CTCs). This process, honed over three decades, has had a deep effect on the organizational culture and the training mind-set of the military. Through the CTC process, the post-Vietnam generations learned to learn.

The idea for such an immense practice ground for tanks to fight tanks in large formations grew from the experience of Navy and Air Force aviation.³ During the Vietnam War, Navy and Air Force leaders learned, through statistical analysis of air combat, that compared to the seasoned combat pilots of the Korean War, American pilots in Vietnam were getting shot down at unacceptable rates. If young pilots could somehow learn what the combat veterans knew *before* their first combat sortie, surely their chances of survival—and of victory—would improve. Was it possible to pass on the lessons of combat in a "bloodless" training environment?⁴

From this question were born the Navy's Top Gun and the Air Force's Red Flag flight training programs. These were high-tech advanced flight schools where pilots could practice against an experienced "enemy" in air-to-air combat. The "enemy" were American instructor pilots, schooled in Soviet tactics, flying fighter planes with Soviet-type capabilities. Anyone who has seen the popular 1980s movie *Top Gun* or the 2004 IMAX film *Red Flag* will have gotten a glimpse (albeit Hollywood-style) into the realism with which these pilots practiced their craft. Technology was the key ingredient at these schools, as it allowed pilots to simulate shooting each other down during real sorties and to review what had happened (versus what they might have *thought* happened) in a debriefing classroom session once they landed.

The idea was that the physical experience of competing against a skilled enemy in the air would allow pilots to develop "muscle memory" and three-dimensional spatial awareness that could not be learned from a book. The after-action debriefing provided an additional, more analytical layer of understanding that accelerated individual pilots' learning curves. Top Gun's success was reflected in the Navy's Vietnam air-to-air kill ratio improving by a factor of five between 1970 and 1973. (The Air Force's Red Flag was not opened until 1975.) Such statistics got the attention of Army leaders such as General Paul Gorman, deputy chief of staff for training, who worked to convince others of the need to have a similar training facility for ground forces.⁵ Gorman's vision was realized in 1982, with the development of the Army's National Training Center in Fort Irwin, California.

The National Training Center (NTC)

The Army's National Training Center is a massive tract of land, roughly the size of Rhode Island, in the southern California desert. Its mission is

---

3. Kitfield, *Prodigal Soldiers*.
4. Ibid., 163.
5. Ibid.

to train large units, as opposed to individuals or small units, allowing full brigades or multiple battalions to practice with large tank formations against an equally massive enemy, called the "OPFOR" (opposing force). Data from the fight is collected from hundreds of cameras placed throughout the 1,000-square-mile site and mounted on tanks, guns, and other vehicles and equipment. As technology improved, so did data collection.

Today, each combatant wears Multiple Integrated Laser Engagement System equipment (MILES gear) that records his or her "death" when shot by either friend or foe, as well as registering who he or she shoots. As if in a massive game of "laser tag," soldiers and vehicles that are shot are removed from the action and declared "dead." Other instruments utilize Global Positioning System (GPS) satellites to pinpoint and record the exact locations of people and equipment throughout the exercise. Together, these data provide a holistic picture of who did what to whom, when they did it, and where—what the Army calls "the ground truth."[6] This technology allows for a level of realism in training that could not be accomplished at a unit's home station. In addition to monitoring technology, home stations lack the required land area, sophisticated well-equipped OPFOR, and, most critically, an experienced team of "observer controllers" (OCs).

The team of OCs is comprised of experienced soldiers at ranks equivalent to the leaders they are observing. Thus, leaders of platoons, companies, battalions, and brigades have a personal OC assigned to monitor their performance. These OCs physically roam the field during the exercise, often stopping to quiz leaders to see what they are thinking and to analyze (and sometimes assist) their decision-making processes. Throughout the exercise, OCs record their observations, which are combined with the data collected throughout the exercise to comprise the official after-action report. This report becomes the "ground truth" for the unit's performance and a key tool for CTC instructors to track trends and adjust training scenarios for subsequent rotations.[7] The OCs at the NTC present their after-action reports to the unit upon completion of the training and provide "take-home packages" (THPs) to assist the commanders in the training at their home stations.

According to one set of organizational scholars, what makes the Army's learning system most effective is the process of after-action re-

---

6. Gordon Sullivan and Michael V. Harper, *Hope Is Not a Method: What Business Leaders Can Learn from America's Army* (New York: Random House, 1996).

7. Personal site visit by the author to the Joint Readiness Training Center. Multiple interviews with OCs and scenario developers, March 2004–March 2005.

view (AAR), with its unique "emphasis of focusing on learning during and immediately after an event and applying what is learned as quickly as possible back into action."[8] Everyone is encouraged to participate by generating unit after-action reports after training and after real-life events and by studying after-action reports from other units in preparation for an exercise or deployment. In this way, "AARs become a process of continuous learning and improvement."[9] Because after-action reports are generated and utilized at all levels of the Army, the systematic focus on learning from action has become a modus operandi for the institution.

Noting the ability of troops in Haiti to make incremental adjustments to their procedures based on information gleaned from their own unit's recent AARs, organizational learning theorist Lloyd Baird claimed, "All of this was learned because the troops on the ground were taught to stop and reflect about what they were trying to accomplish, what they had learned, and how to improve each step of the way."[10] Although they did not "do it right the first time," this unit's systematic recording of events via after-action reports allowed each subsequent attempt to show improved results. The next unit charged with collecting weapons from a Haitian community could literally skip the on-the-job "learning curve" and apply the preferred solution. Or, as Army lessons-learned analysts claim, a unit studying lessons from those who have gone before has a better chance to "do it right the first time."[11] Of course, this last step presupposes that the next unit has access to this information. This is where the Center for Army Lessons Learned comes in.

The Center for Army Lessons Learned

Spawned by the National Training Center as a way to institutionalize the lessons being captured there, the Center for Army Lessons Learned remains a key element in the Army's learning system. The impetus for the center was the recognition that Army units were repeating the same mistakes during NTC exercises over and over again. Although individual units reported a positive learning experience from their NTC rotation,

---

8. Lloyd Baird, John C. Henderson, and Stephanie Watts, "Learning from Action: An Analysis of the Center for Army Lessons Learned (CALL)," *Human Resource Management* 36, no. 4 (Winter 1997): 387.
9. Ibid.
10. Ibid.
11. Center for Army Lessons Learned, "Foreword," *Combat Training Center Quarterly Bulletin*, fourth quarter (2002).

the Army as a whole was not demonstrating improvement over time. Thus, the promise of the NTC—the improved warfighting ability of the U.S. Army—was not being realized. The NTC program, which focused on battalion- and brigade-level training, was expected to go beyond individual soldiering skills; it was to generate institutional learning. Army leaders knew that if the NTC did not demonstrate positive Army-wide results soon, people (particularly congresspeople) were bound to notice.

Operating the NTC had cost taxpayers between $62 million and $90 million a year in its first four years. Additionally, units scheduled for an NTC exercise could expect to spend an additional $4 million to $6 million from their unit funds to finance their particular deployment.[12] Not surprisingly, no sooner had the NTC been established than Congress wanted to know what it was getting for its investment. A GAO report in 1986, *Army Training: National Training Center's Potential Has Not Been Realized,* observed that the Army was identifying many problem areas during NTC exercises through their after-action reports but did not have an adequate system for analyzing the causes of and solutions to these deficiencies.[13] The GAO believed that without a system to track and target trends, the NTC would fail to realize its full potential. To remedy this shortfall, the GAO noted approvingly that the Army had already recently established a system for tracking NTC trends and for collecting and disseminating lessons learned. This new system was the Center for Army Lessons Learned (CALL).

The idea for CALL began in 1984, when the commander of the NTC Operations Group, Colonel Wesley Clark, wrote a white paper outlining the need for a centralized CTC lessons-learned program.[14] In the paper, he suggested the establishment of an independent center comprised of disinterested analysts/subject matter experts (SMEs) whose job it would be to capture and disseminate these lessons. The center would centralize the greater Army AAR/lessons-learned process by acting as a central depository for such information. It would also have a branch dedicated to lessons learned at the CTCs. A primary function of this branch would be to track trends in training deficiencies so they could be targeted for improvement in scenario development and subsequent rotations.[15]

---

12. General Accounting Office, "Army Training: National Training Center's Potential Has Not Been Realized (GAO/NSIAD 86-130)" (Washington, DC, July 23, 1986).
13. Ibid.
14. Colonel Larry Saul, USA, Director, Center for Army Lessons Learned, personal interview by author, March 8, 2004, Fort Leavenworth, KS.
15. George Mordica, numerous personal interviews by author, March 2004–May 2005, Center for Army Lessons Learned, Fort Leavenworth, KS, and via telephone.

By the time the GAO report was published two years later, the Army had already tested their data collection procedures and identified the initial cadre of officers and analysts for CALL.[16] The GAO report noted, "The CALL system is designed to provide needed structure to the Army's subjective NTC assessment efforts, as well as a means to initiate problem solutions. We believe the systematic data collection aspects of CALL are critical to a sound lessons-learned program."[17] Although, at the time, the GAO report seemed to be a scathing criticism of the NTC program, the congressional spotlight on the CALL methodology as a way to address the identified problems invigorated the fledgling CALL project. With a high-level mandate and, more important, an injection of additional Army funds, CALL was off and running.[18] By the late 1990s, the GAO would identify the Army's Lessons Learned Program as a model to be emulated by the other services, the joint community, and the combatant commands.[19] Moreover, independent theorists on management information systems would recognize CALL as a "model" for organizational learning to be studied by academics and the civilian sector.[20]

As the central depository for all the Army's unit after-action reports, CALL is responsible for managing the greater AAR process. This process transcends the production of after-action reports in that it focuses on collection, analysis, synthesis, and dissemination processes as well. Because CALL analysts are removed from the event and because they collect after-action reports from other units (as well as collecting data in other ways described below), they are uniquely situated to cull broader themes, lessons, and trends. Still, as a small shop of less than 15 people, their ability to capture and analyze was limited. Over time, CALL has seen its job less as one of analysis and more as a conduit through which information flows.

In 1993, CALL's mandate expanded beyond the training environment to real-life operations, with a focus on short-, mid-, and long-term

---

16. Marvin Decker, personal interview by author, March 8, 2004, Center for Army Lessons Learned, Fort Leavenworth, KS; General Accounting Office, "Army Training." The four members of the initial cadre were Mr. Marvin Decker, Dr. Lon Siglie, Mr. Kent Harmon, and Mr. Dan Nolan.

17. General Accounting Office, "Army Training," 23.

18. Mordica, interviews.

19. General Accounting Office, "Military Training: Potential to Use Lessons Learned to Avoid Past Mistakes Is Largely Untapped (GAO/NSIAD 95-152)" (Washington, DC, August 1995).

20. Baird, Henderson, and Watts, "Learning from Action."

learning.²¹ In addition to the concerted effort to generate long-term organizational learning focused on "tactics, techniques, and procedures" (TTPs) and doctrine at the training centers, CALL methodology allows for "just-in-time" learning at the tactical and operational levels, even during ongoing real-life military operations. Lessons from the field are collected, analyzed, and disseminated both to the schools to improve training and directly to commanders to inform their current decisions. This has the added benefit of demonstrating quick results, ensuring Army-wide support for CALL collection efforts. Commanders in the field who value the products generated by CALL are thus more willing to cooperate with the center's "collection teams" as they observe units in action.

Since its establishment as a four-man shop in 1985, CALL has grown in size and influence. Today, its small core cadre serving as coordinators and analysts at the modest office in Fort Leavenworth belies the reach and influence the institution has had. Because the CALL system taps into the existing expertise throughout the Army to collect information and lessons learned, the CALL "system" is actually an Army-wide project. The CALL methodology, anchored by the AAR system, is used at all levels of the Army and has generated an Army-wide cultural mind-set focused on learning from experience. Not without its flaws, the system provides a model for how to generate knowledge and promote organizational learning in a large operational organization. Similar programs have been developed in other services. At the joint level, the U.S. Joint Forces Command runs the Joint Center for Lessons Learned. All of these programs are modeled on the CALL method and have leveraged improved technology for information dissemination.

The CALL Method

Three elements comprise the core CALL system: data collection, analysis, and dissemination. For both real-life operations and at the CTCs, data are collected formally and informally. In the most formal method, CALL forms data collection teams known as "combined arms assessment teams" (CAATs), comprised of SMEs tapped from throughout the Army. In practice, most CAAT members come from the TRADOC (Training and Doctrine Command) proponent schools. Once team members are identified, a team chief is chosen, usually a colonel from outside CALL, who works closely with an operations officer and an operations NCO

---

21. Saul, interview.

from CALL. Team members are chosen for their branch expertise (infantry, armor, artillery, etc.) as well as for their personal experience and connections to the particular unit being observed. Composition of these teams varies depending on the requirements of the data collection mission. For example, a unit deploying to Iraq to investigate civil-military operations at the battalion level might include linguists, civil-affairs officers, engineers, and so on.

Although CALL analysts often have ideas about where and when teams should be sent, CALL has only limited funding to generate CAATs. The TRADOC commander often approves CALL's use of these funds for a certain operation or exercise; but usually, a CAAT is requested (and funded) by another unit. In practice, CALL analysts often find ways to ensure a team is sent to a particular operation. This was the case for Task Force Hawk in Kosovo. Recognizing that there were key logistical and operational lessons to be learned from the operations there, CALL members called European Command (EUCOM) headquarters to suggest they submit a formal request to TRADOC for a CALL team.[22] More commonly, commanders in the field often submit a request to the TRADOC commander for a team to accompany their unit either in a real-life operation or for their rotation to a CTC.

At the CTCs, data collection occurs through the identification of trends during training rotations. By Army regulation, the CTC commanders are required to submit trend reports to CALL following each rotation and again semiannually. CALL maintains satellite liaison offices at the CTCs, each of which is led by an Army officer and staffed by two to four civilian SMEs. These small teams interface with the OCs, who also double as SMEs for data collection. For data collection during either a real-world operation or a CTC rotation, the CAAT mission is bounded by specific questions the sponsor hopes to address. CALL personnel work closely with the CAAT to achieve the objectives.

CALL plays an important role in the CAAT process from start to finish. To prepare the team for deployment, CALL hosts a seven- to ten-day collection workshop to introduce team members to the CALL methodology and to devise a "collection plan" for the visit. The requesting commander often has specific questions he wants the team to address, which are factored into the composition of the team and the locations team members are sent.[23] At the end of the workshop, the team will have a list of issues about which they are to collect. These lists are often

---

22. Decker, interview.
23. Mordica, interviews.

quite ambitious, so it is not uncommon for a team to only collect a portion of the items on the list. Following the workshop, team members physically travel to the operation or exercise to collect observations. They interview participants and identify problems as well as innovations. Critical information may be passed immediately back to CALL for dissemination to the training centers and the rest of the Army even while the team is still deployed. Upon return, the team meets to record their observations and lessons learned.

For each item, team members record observations using "CALL-COM" software, on which they are trained at the workshop prior to departure. Each observation is recorded in a standard format, including a one-sentence "observation," followed by a more lengthy "discussion" and ending with either a "techniques and procedures" (if they are at a CTC) or a "recommendation" (if they are in a real-life operation).[24] The techniques or recommendations are informed by the observer's own knowledge and experience as an SME. Each observation can identify weaknesses or problems ("needs emphasis"), as well as innovative solutions ("positive performance"). In order to record an event, observers must be present when the event occurs. This collection of three-part "observations" becomes the raw data for the CAAT's efforts, from which CALL analysts generate various products, described below.

In addition to collecting data through observation at CTC and real-world operations, CALL also solicits data from individual soldiers themselves. This is a more indirect, less formal process that often highlights problems or lessons CALL had not thought to observe. CALL encourages the active participation of Army members at all levels and actively solicits information from soldiers through messages placed in publications. Any Army member can submit information to CALL at anytime. For example, the foreword to the *CTC Quarterly Bulletin* states,

> If you or your unit have a "lesson" that could help other units do it right the first time, send it to us. Don't worry about how polished your "article" is. CALL can take care of the editing, format, and layout. We just want the raw material that can be packaged and then shared with everyone.[25]

The "packaging and sharing" of this "raw material" is a key role for CALL. "Packaging" requires analysis and synthesis of data, while "sharing" requires a system for dissemination of the products.

---

24. Decker, interview.
25. Center for Army Lessons Learned, *Operations Other than War*, vol. 4, *Peace Operations* (Fort Leavenworth, KS: U.S. Army Command and General Staff College, 1993).

The Products: CALL Trends Analyses, Dissemination, and Customer Service

CALL disseminates information to the Army in multiple formats, including electronic and print media. To highlight training issues, CALL publishes the *CTC Quarterly Bulletin,* containing training-related articles written "by soldiers for soldiers." This magazine-size publication (approximately 50 to 60 pages) is a product of the "raw data" request quoted above. CALL chooses from the many submissions each quarter with the goal to reflect a cross section of Army interests. Also comprised of soldier-submitted articles is the periodic *CALL Newsletter.* With no set publication cycle, this newsletter is published to highlight a specific topic, such as urban operations, artillery procedures, and so on. These publications are also populated with articles submitted by soldiers. Other "raw data" soldier submissions may appear in the bimonthly publication *News from the Front!* or may be posted on the CALL website under "Training Techniques." In addition to these periodic training-focused publications, CALL provides in-depth publications on lessons learned for real-world operations, as well as topical handbooks on tactical subjects like convoy operations, urban warfare, and so on. The handbooks are small enough to fit in a soldier's fatigue pocket.

One of the more popular handbooks, the *Stability and Support Operations Handbook,* was produced in July 2003 in just one month, in an unusually high-level effort, and highlighted a gap in tactical doctrine relevant to current operations. In June 2003, the Israeli Army's chief of staff, Lieutenant General Moshe Ya'alon, invited the U.S. Army chief of staff, General Shinseki, to send a team to observe Israeli tactics on counterterrorism, sniper operations, convoy operations, and so on. CALL sent a CAAT team of eight Army officers, four marines, and two Homeland Security officials, who flew to Israel for a 10-day observation. Their after-action report was posted on CALL's website before the team even returned, and the book was published within the month. This little gray booklet was widely disseminated and became ubiquitous in the training areas. It was also available for download as a PDF file from the CALL website.[26]

The CALL website, now the primary method of dissemination for CALL products, is a password-protected online medium through which military personnel can access after-action reports, trends, special reports,

---

26. Mordica, interviews.

and the various publications listed above.[27] As one senior CALL analyst explained, "everything one needs to prepare a unit to go is on this website." The pure volume of information available on the site is often a point of criticism however, as some find that the search mechanism and the overall organization of the site create a situation of "information overload."[28] One way around this problem is to use the site's "request for information" (RFI) page.

The RFI system works by allowing commanders to e-mail or phone the CALL analysts, who then compile all relevant information onto a custom-made CD or, if possible, for e-mail transfer. For instance, when Captain Alex Mentis, a company commander in the 326th Battalion of the First Brigade of the 101st Airborne Division, learned in the fall 2004 that he would have to train his noninfantry unit to deploy for stability operations in Iraq as part of an infantry brigade, CALL was the second place he turned for advice.[29] (The first thing he did was to ask his colleagues.) With the help of CALL analysts who sent him information on tactics, techniques, and procedures being collected in the field, as well as the latest scenario information for stability operations training, Captain Mentis was able to craft a training program for his company that became the standard for the 326th Battalion.[30]

The story of Captain Mentis and his three fellow company commanders in the 326th Battalion is a reflection of the Army's controversial transformation to a "Modular Force."[31] This battalion was one of the first "units of action"—a new aspect of the Army's modularity plan that incorporates companies providing combat support into an infantry brigade. The 326th Battalion was led by Lieutenant Colonel Smith, an Army engineer, and was part of the First Brigade of the 101st Airborne Division, commanded by Colonel David Gray. The battalion contained

---

27. http://call.army.mil/.
28. Numerous interviews with junior and field grade officers throughout the Army and Marine Corps.
29. Captain Alex Mentis, USA, personal interview by author, October 13, 2004, Fort Campbell, KY.
30. Captain Wess Dumas, USA, personal interview by author, October 13, 2004, Fort Campbell, KY; Captain Chip Hann, USA, personal interview by author, October 13, 2004, Fort Campbell, KY; Captain Steve Walters, USA, personal interview by author, October 13, 2004, Fort Campbell, KY. These three captains, in addition to Captain Mentis, were the four company commanders for the 326th Battalion in the 101st Airborne.
31. Army news release, "Army Announces Fy05 and Fy06 Modular Brigade Force Structure Decisions," *Army Public Affairs,* July 23, 2004. "Modularity" refers to the Army's attempt to make itself more rapidly deployable. See also Tammy Schultz, "Ten Years Each Week: The Debate over American Force Structure for Winning the Peace?" (PhD diss., Georgetown University, 2005).

four companies, each comprised of noninfantry troops from various support branches and led by captains from the signal, medical, military intelligence, and engineering branches. Thus, in preparation for their deployment to Iraq in the fall of 2005, these four noninfantry captains and their commander from the engineering branch were charged with training noninfantry troops for infantry-type tactics, techniques, and procedures such as combat patrolling, fire and maneuver, and convoy operations in hostile terrain. Although these captains lamented the lack of formal education they had received during their seven-year careers, their ability to prepare for this challenging mission was greatly enhanced by the Army's dynamic tactical learning system, described here. This system included the products and services provided by the Center for Army Lessons Learned, which were in turn being fed new information daily from Iraq, Afghanistan, and elsewhere around the world.

From Experiential to Organizational Learning through the CTC Model

The CTC and CALL processes allow for all four elements of David Kolb's familiar experiential learning cycle (fig. 3): (1) concrete experience, (2) observation and reflection, (3) forming abstract concepts, and (4) testing in new situations.[32] Following the mock battle scenario corresponding to step 1, steps 2 and 3 require a concrete picture of what happened and why, as well as an analysis of how to do things better next time (step 4). Steps 2 and 3 are folded into the AAR and debriefings at the CTCs, while step 4 is eventually played out in training at home stations or upon deployment to the real theater of operations.

During the CTC exercise, the critical learning event is the AAR debriefing, during which OCs and other officers question leader decisions and suggest alternatives. This corresponds to steps 2 and 3 in the cycle, allowing participants to review the events and form a mental assessment of what went right and what went wrong. Just as when a football team reviews the game tape the day after the lost match, all the sensing equipment on the "battlefield" and the ability to identify exactly who did what and where leave little room for debate over the course of events. In this environment, there is little room for arrogance as well. As one observer described the AAR environment, "the after-action reviews began to take on the feel of a group therapy session for a somewhat dysfunctional family, with junior officers increasingly questioning the decisions and ac-

---

32. D. A. Kolb, *Experiential Learning* (Englewood Cliffs, NJ: Prentice-Hall, 1984).

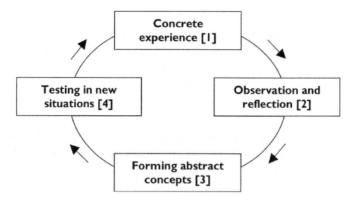

Fig. 3. Experiential learning cycle. (Data from D. A. Kolb, *Experiential Learning* [Englewood Cliffs, NJ: Prentice-Hall, 1984].)

tions of their superiors that had just gotten them destroyed on the battlefield."[33] With the stakes as high as potential defeat on the battlefield, truth becomes increasingly more important than ego.

The learning environment promoted by the CTC model had a profound effect on the Army. James Kitfield observes,

> Though they did not yet realize it, that willingness of junior officers to openly question their superiors, and of superior officers to admit mistakes in front of their subordinates, was beginning to fundamentally change the culture of the Army. An organization that would once have considered such behavior little short of insubordination began to encourage self-criticism in an effort to get at the truth.[34]

Moreover, the importance of critical after-the-fact reflection on one's own actions and the actions of one's unit began to be instilled in soldiers at all ranks. When OCs shadow leaders during the exercise, asking them what they are thinking and why, soldiers learn to reflect on their internal decision-making processes and on the performance of their unit.

The scenarios practiced at the CTCs are derived from formal doctrine and real-life operations. The scenario developers and OCs at the centers have shelves of well-worn field manuals from which they are constantly pulling references to incorporate into scenarios and to discuss in the AARs. Thus, as General Kevin Byrnes, the former commander of the Army's Training and Doctrine Command, claimed, the CTCs function as

---

33. Kitfield, *Prodigal Soldiers*.
34. Ibid., 311.

"injection points" for new knowledge, new doctrine, and even Army culture.³⁵ Although officers may not have read the latest formal doctrinal manuals, they become familiar with their contents through practice at the centers. As one veteran explained the link, if the learning system is functioning properly, doctrine becomes like the Bible in that "few people read it, but most are familiar with the Ten Commandments."³⁶

It is important to recognize that this process was no accident. Army leaders were familiar with the new theories of organizational learning as they developed and refined the CTC program. As General Gordon Sullivan writes, "the AAR system was the key to turning the corner and institutionalizing organizational learning." Still, he describes a "learning organization" as an elusive concept, "something you aspire to always 'becoming,' yet never truly 'being.'" Later, as chief of staff of the Army in the early 1990s, General Sullivan would actively consult organizational learning theorists such as Peter Senge to help improve the Army's ability to adapt to what he knew was a rapidly changing post–Cold War environment.³⁷

In sum, the National Training Center was a concerted effort by the Army leadership to promote "learning through doing" at the unit level. This realistic scenario-based training allowed the various elements of an armored brigade to practice a choreographed battle plan against an opposing force. While this innovation improved the readiness of the big army to conduct large land war against the Soviets, it did little to address requirements at the lower end of the spectrum. For this, light forces and Special Operations Command pushed for the creation of a separate training center, the Joint Readiness Training Center.

The Joint Readiness Training Center: A Place to Call Their Own

Writing of his battalion's experience when confronted with a hostile mob of armed Haitian civilians in November 1995, Lieutenant Colonel Dan Bolger (later promoted to general) claimed,

> The way [Staff Sergeant Brian Cagle] reacted, and the calm discipline shown by this NCO and his men came from training like that conducted at the Joint Readiness Training Center. Indeed, the reason a solitary rifle squad did not hesitate to quell an ugly mob came from

---

35. General Kevin Byrnes, USA, Commander Training and Education Command, personal interview by author, February 7, 2005, Fort Monroe, VA.
36. Joseph Collins, personal interview by author, April 15, 2004.
37. General Gordon Sullivan, USA, personal interview by author, February 25, 2005, Arlington, VA.

the confidence bred in the swamps, forests, and sham villages of JRTC.[38]

At the time of this unit's deployment to Haiti, little definitive doctrine existed for infantry forces facing such a scenario, and it was not the type of thing that would be practiced at the NTC. According to Bolger, the "muscle memory" developed at the JRTC the previous year had made Staff Sergeant Cagle's reaction seem like "common sense" to him. The fact that an entire quick reaction force of Bangladeshi peacekeepers stood by and watched in awe as the small American squad quelled the crowd, handcuffed angry machete-waving Haitians, and redeployed without a scratch suggests, however, that it was not. How was this possible? What was the JRTC, and what exactly were infantry forces being taught there?

The JRTC Story

The Joint Readiness Training Center was spawned from four overlapping post–Vietnam War initiatives: the NTC, CALL, the resuscitation of the Special Operations Forces, and the development of the Light Infantry. Even before Somalia, the widely acclaimed success of the NTC training model developed and honed in the 1980s led to the realization that the Army's various "light" combat forces also needed a place to train. Having demonstrated the value of realistic training on a large unit level, the institution could not politically justify only applying this model to part of its forces. Doing so would mean an Army only partly "ready" to carry out its mission.

An NTC rotation was considered the pinnacle event of a battalion or brigade commander's tour of duty, and an assignment as a trainer at the NTC was considered career-enhancing (General Wesley Clark, who ultimately became the supreme allied commander in Europe, spent the Gulf War running the NTC rather than fighting in the Kuwaiti desert), but others criticized the scenarios as irrelevant. One officer explained,

> NTC has never reflected the most likely battles the U.S. Army has fought and will fight . . . Even at the height of the cold war, many soldiers noticed a disconnect between the straightforward, blitzkrieg nature of the Army's premier unit training event at Fort Irwin and the muddled, frustrating character of the unremitting succession of con-

---

38. Daniel P. Bolger, *The Battle for Hunger Hill: The 1st Battalion, 327th Infantry Regiment at the Joint Readiness Training Center* (Novato, CA: Presidio, 1997).

fused, dirty little operations that kept occupying Americans between 1945 and the fall of the Soviet Union.[39]

Advocates of light forces recognized that since 1945, the Army had spent only four days fighting the type of large tank-on-tank battles practiced at the NTC.

Light Infantry and Special Operations

The drive to revitalize the "Hollow Force" was focused primarily on "big war." Accordingly, all power went to the engines of change as focused on defense of the Fulda Gap and war with the Soviet Union. The new Air-Land Battle Doctrine, which was written by armor officers and heavily focused on tank battles, would be the military's raison d'être.[40] In this atmosphere, small wars, COIN, and other low-intensity operations were virtually ignored by the conventional forces. Even the special counterinsurgency-oriented forces from the Kennedy era withered and nearly died. One military historian concluded,

> By the end of the 1970s, US special operations forces (SOF) were *caput mortuum*. Army special forces had been gutted, Navy special warfare had fared little better, and Air Force special operations forces (AFSOF) had barely survived a concerted attempt to relegate them completely to the Reserves.[41]

By the mid-1980s, efforts began to address this shortfall through two separate initiatives, the development of the Light Infantry and the creation of the Special Operations Command.

In 1984, following on the heels of various experiments conducted with light forces by General "Shy" Meyer, General John A. Wickham, Army Chief of Staff from 1983 to 1987, issued a white paper outlining a more operational Light Infantry concept.[42] In strong language that challenged the big-war status quo, Wickham claimed,

> Army leadership is convinced, based on careful examination of studies which postulate the kind of world in which we will be living and

---

39. Ibid., 11–12.
40. Killebrew, interview.
41. http://www.airpower.maxwell.af.mil/airchronicles/apj/spr97/johnson.html.
42. General John A. Wickham, interview with Chief of Staff of the Army General John A. Wickham, Jr., Army Chief of Staff, *Armed Forces Journal* (September 1985); see also "Introductory Letter: White Paper 1984, Light Infantry Division," in General John A. Wickham, Jr., Army Chief of Staff, *Collected Works of the Thirteenth Chief of Staff of the United States Army* (Washington, DC: Center for Military History, 1984).

the nature of conflict we can expect to face, that an important need exists for highly trained, rapidly deployable light forces. The British action in the Falkland Islands, Israeli operations in Lebanon, and our recent success in Grenada confirm that credible forces do not always have to be heavy forces.[43]

The Light Infantry would be part of the new "Army of Excellence" initiatives, and its mission, according to Wickham, would be "clearly defined." They would "function in the low- to mid-intensity environment, to get to crisis areas rapidly, either to deter hostilities or to influence them to our advantage."[44] Echoing General Meyer's emphasis on Europe, Wickham also pointed out, "There is a fair amount of Central Europe that is forested, that is wetland, that is urban sprawl with substantial obstacles." Thus, light divisions were marketed as necessary for the "big fight" against the Soviet Union as well. This additional truth helped sell the concept to Army skeptics.

Meanwhile, poor military performance in Grenada and the failed attempt to rescue hostages in Iran at Desert One got the attention of lawmakers. Congressional action in the mid-1980s, culminating in the Goldwater-Nichols Act of 1986, was clearly directed at the failures in command and control among the services during the Grenada invasion. Additionally, through the 1987 Nunn-Cohen Amendment, a lack of capability at the "low-intensity" end of the spectrum was targeted, and the Special Operations Forces (SOF) were slowly brought back to life.

Three critical changes were stipulated in the Nunn-Cohen Amendment: the creation of the Special Operations Command (SOCOM), the establishment of a new Pentagon position for an assistant secretary of defense for special operations and low-intensity conflict, and a dedicated congressional funding line in the DoD budget (Major Force Program 11). Under the direction of a four-star general, SOCOM was directed to develop appropriate doctrine, education, and training to ensure the professional development of SOF personnel and the overall readiness of the new command. Thus, the Nunn-Cohen Amendment laid the foundation for the resurrection of the Special Forces and the incubation of doctrine for MOOTW and for counterinsurgency and stability operations.[45] These two communities together lobbied for enhanced training facilities on par with those available to the heavy forces at the NTC.

---

43. Wickham, "Introductory Letter."
44. Wickham, *Armed Forces Journal* interview.
45. Lieutenant Colonel Wray Johnson, "Whither Aviation Foreign Internal Defense?" *Aerospace Power Journal* (Spring 1997); Susan L. Marquis, *Unconventional Warfare: Rebuilding US Special Operations Forces* (Washington, DC: Brookings Institution Press, 2003).

The JRTC Begins

In a nod toward addressing the "dirty little operations" of the 1980s, the JRTC opened at its temporary home in Fort Chaffee, Arkansas, in 1987. With the Soviet threat still very real, however, the problem of the Light Infantry and the Special Forces in irregular or urban operations was considered a lesser priority. Indeed, the JRTC did not get its own home, in Fort Polk, Louisiana, until 1993 and did not get an NTC-style instrumentation system until 1996, a full 14 years after the NTC was built. Although the realities of the post–Cold War world would eventually have both the NTC and the Combat Maneuver Training Center (CMTC) tapping into many of the "irregular" lessons learned at the JRTC, the pearl (and the priority) of the Army was the NTC, at its inception and for over a decade and a half thereafter.

In contrast to the NTC's focus on armor, the JRTC was specifically designed to reflect the conditions under which the Army's 42 "dismounted" infantry battalions and Special Forces would likely fight.[46] In addition to the Special Forces, these light battalions fall into four categories: Airborne, Air Assault, Rangers, and Light Infantry. Together, they collectively comprise the combat part of the Army known as "boots on the ground." Recently, part of the Light Infantry force has been converted to the new Stryker Brigade Combat Team, characterized by the use of the new medium-weight wheeled combat vehicle called "the Stryker."

The Light Infantry and Special Forces created a critical mass of the types of forces that would together prove uniquely suited to the problems of the post–Cold War era. Although conventional wisdom is that the 1990s military was a Cold War force focused on the big-war paradigm, the development of these light forces and the concurrent resurrection of the Special Forces (even though accomplished as an apparent afterthought in the mid-1980s) planted a powerful seed that provided the raw material for stability operations and COIN learning into the next decade. The JRTC and the myriad real-world stability operations of the 1990s provided the classroom.

It is no mistake that these light forces have deployed to every significant stability operation starting with Panama and continuing to Iraq and Afghanistan today. Lessons learned by the 82nd Airborne in Panama and by the Rangers and 10th Mountain Division troops in Somalia were reflected in training scenarios developed at the JRTC even

---

46. Since 2005, reorganization efforts have begun to adjust the numbers of battalions in many of these divisions and increase the overall size of the Special Forces.

before any new material had been recorded into formal doctrine. In 1994, over 40 percent of the troops sent to Haiti from the 10th Mountain Division had also served in Somalia.[47] The experiential learning by this small community of light forces was critical in this period. By the time subsequent light forces deployed to Haiti and the Balkans, lessons had filtered through the JRTC and via informal networks to the next generation of "boots on the ground."

The JRTC Experience

Capitalizing on the success at the NTC, Fort Polk's JRTC trainers employ the same experiential learning–oriented philosophy as their counterparts in the California desert. Like the NTC, the JRTC employs a team of OCs, an OPFOR, AARs, and THPs.[48] Since the mid-1990s, it also has much of the same instrumentation used at the NTC, including MILES gear and the Blue Force Tracker, which enables commanders to see where all the "friendly" forces are via a computer screen mounted in their Humvee. These systems allow for data collection throughout the battle space and for the presentation of the "ground truth" in the AAR. Thus, "learning through doing" and the development of "muscle memory" through practice are reinforced with rigorous after-action analysis, in the same ways and through the use of the same types of instrumentation as at the NTC.

Despite similarities in learning theory and instrumentation, the JRTC differs significantly from the NTC in other important ways. Instead of a wide-open desert, Fort Polk is carved out of a swampy, wooded jungle in Louisiana. Instead of tanks firing in an open field, the JRTC battlefield is a series of urban towns and rural villages, complete with civilians on the battlefield (COBs) who role-play as everything from NGOs and UN officials to refugees, media, and other interagency actors. More important than the JRTC's physical characteristics, however, are the different scenarios practiced and OPFOR tactics used. The JRTC scenarios incorporate a combination of insurgent and regular enemy forces, which together challenge the intervening American brigade.

The JRTC scenario is designed for modern-day complex contingencies. In addition to an NTC-type OPFOR that utilizes more traditional

---

47. Walter E. Kretchik, Robert F. Baumann, and John T. Fishel, *Invasion, Intervention, "Intervasion": A Concise History of the U.S. Army in Operation Uphold Democracy* (Fort Leavenworth, KS: U.S. Army Command and General Staff College Press, 1998), http://www.globalsecurity.org/military/library/report/1998/kretchik.htm.

48. Observations on JRTC are based on personal site visit by the author, March 2004.

warfighting doctrine and tactics, the JRTC scenarios add irregulars and terrorists who employ guerrilla tactics among a vulnerable, needy, and sometimes hostile civilian population. Although today's scenarios are based on Iraq and Afghanistan and on those in Bosnia in the 1990s, the original 12-day JRTC scenario reflected the "dirty little war" scenarios of the 1980s. American forces would be sent to the fictitious island of Aragon to support the weak, third-world nation of Cortina. Cortina would be struggling with the Cortina Liberation Front (CLF), a Marxist guerrilla group assisted by Cortina's well-armed neighbor to the east, the Peoples Democratic Republic of Atlantica. The CLF, played by the experienced OPFOR, presented a wily menace throughout the rotation, as the "low-intensity," low-tech, insurgent force. By the time the "real" enemy, Atlantica's conventional armored force, invades from the East about halfway through the scenario, the Americans were practically relieved to be faced with a "normal" fight.[49]

In this traditional scenario, the rotation usually ended with a final siege against the objective: the Cortinian town of Shugart-Gordon. The town is named after two Medal of Honor winners, Sergeant First Class Randall Shugart and Master Sergeant Gary Gordon, who died rescuing a downed pilot in Mogadishu, Somalia. The significance of the name of this town is not lost on soldiers training at the JRTC. Indeed, among the many lessons that the JRTC scenario developers have taken from the experience in Somalia is the difficulty of fighting in an urban environment against fierce insurgents and "irregular" forces. One JRTC trainer claims, "We don't want units thinking that seizing control of Shugart-Gordon was easy. We want them to know that as much as they think they know about urban warfare, dilemmas such as sleep deprivation, lack of time to prepare, can be life threatening."[50] This theme that "low-intensity" conflict in modern urban environments was not to be taken lightly, or as something "lesser than" war, would recur throughout the decade as troops trained for back-to-back rotations to the Balkans.

The JRTC and the Original Focus on MOOTW

The importance of the JRTC model became increasingly apparent starting in the early 1990s and especially following the episode in Somalia. Although the lesson of Somalia for many military and civilian leaders was

---

49. Bolger, *The Battle for Hunger Hill*.
50. Major Perry Beissel quoted in Staff Sergeant Marcia Triggs, "Fighting in Urban Terrain Challenging, Not Impossible," *Army News Service*, November 27, 2002.

to steer clear of such complex, low-intensity missions, others knew they were unlikely to be avoided. Again, conventional wisdom reflecting the end of the Vietnam War would have us believe that the military would have ignored the challenges presented by Somalia. Although there were those who were indeed afflicted by what Richard Holbrooke called the "Vietmalia syndrome"[51] and who searched for ways and means to avoid such missions, there were other influential leaders who set out, instead, to find ways to perform them better. One of these leaders was Army chief of staff General Gordon Sullivan.

In August 1992, just months before troops were sent to Somalia, General Sullivan directed the CMTC in Germany to incorporate MOOTW into its training scenarios. For the first time, MOOTW-oriented METLs such as caring for refugees and conducting patrols were developed for units training at the center. This continued through 1993, when the new FM 100-5 *Operations* (discussed in chapter 5) was released. At General Sullivan's urging, the manual included, for the first time, an entire chapter on MOOTW. Also in 1993, the JRTC was reopened at its new, improved location in Fort Polk. By 1994, with the events of Somalia fresh in mind and conflicts in Haiti and the Balkans looming, Sullivan continued to emphasize MOOTW-oriented doctrine, education, and training. In that year, his new Peacekeeping Institute (PKI) opened at the Army War College, and peace operations training (then considered a more specific subset of MOOTW) was being conducted at both the JRTC and the CMTC. To highlight his commitment to this training, he personally observed a JRTC rotation in September of that year.[52]

In the summer of 1994, two training events marked the new emphasis on peace operations training. In June, the First Armored Division conducted a predeployment exercise at the CMTC specifically designed to emulate the environment they might face in the Balkans. The scenario was based on a fictional country of Danubia, in which the unit was required to set up a "zone of separation" between belligerents and then keep the peace among three ethnic groups and seven factions. Complicating the task were an indigenous civilian population, the media, and regular and irregular forces.[53] On December 20, 1995, the First Armored Division deployed to Bosnia as the lead unit for IFOR (Imple-

---

51. Richard Holbrooke, *To End a War* (New York: Random House, 1999).
52. Bolger, *The Battle for Hunger Hill;* Sullivan, interview.
53. Department of the Army, *Annual Historical Summaries* (Washington, DC: Center for Military History, 1983–99); Anne W. Chapman, *The Army's Training Revolution: 1973–1990, an Overview* (Fort Monroe, VA, and Washington, DC: Office of the Command Historian, United States Training and Doctrine Command and the Center of Military History, 1994).

mentation Force). Charged with enforcing the Dayton Peace Accords, the division encountered many of the challenges for which it had trained at the CMTC. The country of Danubia would remain a fixture at the CMTC throughout the decade. As troops returned from the Balkans with after-action reports, the scenarios were adjusted accordingly.

The next significant event for peace operations training occurred in Louisiana, when the Second Brigade of the 25th Infantry Division (Light) from Hawaii, commanded by Colonel Charles Swannack (later promoted to general, serving as division commander of the 82nd Airborne in Iraq), conducted the first JRTC exercise oriented toward peace enforcement. Both the 25th Infantry Division's commander, Major General George A. Fisher, and the JRTC commander, Brigadier General Lawson W. Magruder III, oversaw the development of the scenario and the execution of the event. General Magruder had replaced General Fisher as the Fort Polk commander and had also commanded troops in Somalia. Together, these two leaders crafted a scenario based on their personal experiences and knowledge of the capabilities of the JRTC. Although they did not know it at the time, this exercise would serve as an ideal rehearsal for the brigade's deployment to Haiti the following year.

The scenario that the 25th Infantry Division practiced in the late summer of 1994 was based specifically on the lessons learned in Somalia and the conflict brewing in the Balkans and was thus a twist to the "normal" JRTC exercise in low-intensity conflict. Instead of entering the fictitious island of Aragon as an intervening force in support of Cortina, the brigade deployed with the mutual consent of the belligerents and with a UN mandate to enforce the peace. The belligerents of the two fictitious countries had agreed to a cease-fire and had requested a UN intervention force to supervise a contested buffer zone. Unlike a typical JRTC scenario in which there is an identifiable enemy, the American brigade's job was to "separate belligerents, clear a buffer zone, aid in humanitarian assistance (HA), and set the conditions for a relief in place by UN PK [peacekeeping] forces." Other distinguishing features of the scenario were that the brigade would encounter a country "where ethnic strife, civil war and competing insurgencies have caused untold human misery."[54] Thus, instead of ignoring civilians, interacting with the local population and actively trying to ease human suffering were core elements of their job description.

---

54. Colonel Charles H. Swannack, U.S. Army, and Lieutenant Colonel David R. Gray, U.S. Army, "Peace Enforcement Operations," *Military Review* 76, no. 6 (November–December 1997).

Given this complex scenario, the JRTC "battlefield" was teaming with "a bewildering array" of NGOs, insurgents, and refugees. While the OPFOR played the role of insurgents and regular combatants, civilian contractors were hired to play the roles of civilian leaders and refugees. In this particular high-profile exercise, real-life international officials from allied countries acted as UN observers, while American diplomats, Ambassador Robert Oakley and Walter Clarke, served as special presidential envoys. Representatives from real NGOs such as Save the Children, Food for the Hungry, and the International Medical Corps also participated, often presenting some of the greatest challenges to company commanders.

The presence of NGOs in this exercise was an important event. The handful of military officers who had served in Somalia had reported feeling generally unprepared to deal with NGOs.[55] As discussed in chapter 5, General Sullivan, who had visited Somalia in the early phases of the mission, came to the immediate conclusion that "we as an institution do not understand the dimensions of NGOs very well." Likewise, he felt that the Army was viewed by the NGO community as "killers on the battlefield" and thus treated with suspicion. Accordingly, many of his initiatives as chief of staff of the Army, from the establishment of PKI to the promotion of MOOTW and peace operations training at the CTCs, were intended both to help the U.S. military understand NGOs better and to reach out to the NGO community to "improve the Army's image."[56] Throughout the 1990s, this learning extended to the junior officer level both at the training centers and in the field. Thus, while colonels worked with high-level UN officials and participated in negotiations to secure the peace on the fictitious island of Aragon, captains and lieutenants haggled with NGOs over safe areas, refugees, and security procedures.

Writing of the experience in *Military Review*, the 25th Infantry Division's training officer, Lieutenant Colonel David Gray, advised others that in a typical operation of peace enforcement (PE), it would be up to brigade-level leadership to derive concrete mission objectives from murky political guidelines. More specifically, he provided the following advice to his fellow training officers.

> Units assigned PE duties must focus some of their collective training on constabulary tasks, such as setting up roadblocks and checkpoints, patrolling in urban areas and aiding civilian refugees. Most of these tasks can be superimposed over existing drills, tactics, techniques and

---

55. Sullivan, interview.
56. Ibid.

procedures and METL tasks. Others will require ad hoc responses and ingenuity . . . Brigade [task forces] performing PE duties must make certain intellectual adjustments to adapt to this convoluted military environment.

When Lieutenant Colonel Gray and his commander, Colonel Swannack, wrote the article in *Military Review* chronicling the lessons their brigade had learned at the JRTC, they were fully aware that their training exercise was an important event.[57] General John Shalikashvili, chairman of the Joint Chiefs of Staff, had personally visited Fort Polk to observe the exercise. Speaking directly to the division's leaders, the general asserted, "Peace operations are not *instead* of what our business is all about. This is *in addition to*. The world is going to have an awful lot of these operations. This is not going to come to an end when Rwanda is over or Bosnia is over."[58] Moreover, accompanying him to reflect the commitment to U.S.-UN peacekeeping was U.S. ambassador to the United Nations Madeline Albright. This visit by two high-level U.S. officials naturally generated the interest of the media and demonstrated consensus between the Pentagon and the White House. More important, it sent a clear message to troops and detractors. Ambassador Albright told reporters, "The message is that the Army sees peace operations increasingly as part of its mission, not just something that's an ideal in the minds of academicians or multilateralists."[59]

With the support of the chairman of the Joint Chiefs of Staff and the administration, and in light of the conflicts in Somalia, Haiti, and the Balkans, it seemed clear that peacekeeping was to be a core mission for the U.S. Army. Later in 1994, at the JRTC, the 82nd Airborne from Fort Bragg, North Carolina, performed a peace operations exercise, and the Second Armored Cavalry Regiment's Third Squadron from Fort Polk conducted a "mini"-exercise in preparation for deployment to Haiti. For the latter, instead of the faux countries of Cortina and Acadia, the training center was transformed to look "just like Port-au-Prince."

At the time of these initial peace operations exercises, there was very little formal doctrine in circulation. In addition to the brand-new MOOTW chapter in the 1993 FM 100-5, Swannack and Gray cite only FM 100-20 *Military Operations in a Low-Intensity Conflict* and its compan-

---

57. Colonel David Gray, USA, numerous personal interviews by author, 2004–5, Brookings Institution, Washington, DC, and Fort Campbell, KY.
58. "General: Training Key to Worldwide Missions," *Times-Picayune*, August 15, 1994.
59. Bradley Graham, "New Twist for U.S. Troops: Peace Maneuvers," *Washington Post*, August 15, 1994.

ion tactical-level manual FM 7-98 *Operations in a Low-Intensity Conflict.* Mostly, they claim, the unit referenced "evolving doctrinal frameworks" such as a white paper issued by the Army's Infantry School, *The Application of Peace Enforcement (PE) at the Brigade and Battalion Level*,[60] a 1993 pamphlet from the Center for Army Lessons Learned, *Operations other than War*, volume 4, *Peace Operations*,[61] and "various techniques, tactics and procedures packets provided by USAIS [US Army Infantry School] mobile training teams." Gray and Swannack concluded that their exercise "highlighted many issues for further doctrinal study," because "the Army's participation in peace operations and operations other than war will likely remain high in the short term."[62]

Gray and Swannack's prediction in 1994 was validated as the decade progressed. Once it became clear that the mission to Bosnia would last more than the promised year, the Army's learning system began to adapt. Lessons from ongoing operations in Haiti and the Balkans were filtered back into the scenarios at the JRTC via personal experiences, such as the one recorded by Lieutenant Colonel Bolger, and through the more formal process facilitated by the Center for Army Lessons Learned. By the late 1990s, regular scenarios at the CMTC and the JRTC were virtual replicas of Bosnia and Kosovo. This shift toward MOOTW-oriented training at these centers reflected real-life experiences of the next generation of leaders and was an important step toward organizational learning beyond "big war."

In 2003, when the U.S. military found itself confronting a messy counterinsurgency following the swift invasion of Iraq, the system again adapted. The National Training Center in Fort Irwin, which had remained focused throughout the 1990s on scenarios of major combat requiring large maneuvers, rapidly shifted its focus to the exigencies of Iraq. This rapid shift leveraged the hard-learned lessons of the JRTC from the previous decades. Scenario developers from the JRTC quickly transferred their material and techniques to California. Within months, the NTC had hired role players and created the JRTC experience on a larger scale. It is clear that without the work of the JRTC and the efforts of the Light Infantry and SOF throughout the 1990s, the adaptation at the NTC would have been much slower.

---

60. U.S. Army Infantry School Commandant, *White Paper: The Application of Peace Enforcement (PE) at the Brigade and Battalion Level* (Fort Benning, GA, August 31, 1993).
61. Center for Army Lessons Learned, *Operations Other than War*, vol. 4, *Peace Operations* (Fort Leavenworth, KS: U.S. Army Command and General Staff College, 1993).
62. Swannack and Gray, "Peace Enforcement Operations."

Communities of Practice and Other Informal Networks of the 21st Century

Further enhancing this rapid 21st-century adaptation from the field to the training centers was the emergence of technology-enabled communities of practice (CoPs). *Community of practice* is a term coined by learning theorist Etienne Wenger to describe a group (a "community") of professionals who share ideas with the goal of learning from each other and advancing the profession. CoPs can meet anywhere to facilitate the transfer of knowledge, but today's CoPs increasingly meet in cyberspace. In the past few years, online CoPs have proliferated in the military. The most well-known of these is CompanyCommand.com.

CompanyCommand.com was started in the spring of 2000 by a group of officers at West Point. They describe CompanyCommand.com as "a network of company commanders who connect in conversation about relevant content to advance the practice of company command." They share their experiences, discuss and debate them, and reason and learn together. In a top-down, doctrine-driven rigid organization, as is the stereotype of the U.S. Army, such behavior would seem a threat. What if these junior officers are reasoning their way toward poor choices or nondoctrinal solutions?[63] Although there were a few in the organization who felt this way, the official response from the U.S. Army was to provide the leaders of CompanyCommand.com more resources to run their website, encourage them to start another one at the platoon level (Platoon Leader.com), help them publish a book about the their experience, hire an outside organizational learning expert/consultant to assist them, and create an organization for the study of leadership based around their concept (the Center for the Advancement of Leader Development and Organizational Learning) and let them run it.[64] This was all accomplished in a two-year time frame.

Wenger explains the relationship between a CoP and an organization as follows:

> Despite the misgivings of some organizational leaders, community does not compete with hierarchy. Both must exist in an organization,

---

63. Although a few managers at CALL initially expressed concern over CompanyCommand.com, all senior leaders interviewed for this research fully supported the idea, including General Byrnes, the four-star general in charge of the Army's Training and Doctrine Command. Retired Chief of Staff of the Army General Gordon Sullivan formally endorsed the book with laudatory comments on the book's back cover. Sullivan, interview.

64. Nancy M. Dixon et al., *Companycommand: Unleashing the Power of the Army Professions* (West Point, NY: Center for the Advancement of Leader Development and Organizational Learning, 2005).

and they don't subsume each other. They are two sides of the same learning coin, complementing each other to build a knowledge organization. Communities generate learning, and hierarchies put that learning to work to accomplish organizational goals.[65]

Thus the challenge for the U.S. military is not simply to permit these CoPs to function, although that is important, but to somehow harvest what they are producing as a means of generating bottom-up institutional learning. This is especially critical during periods of high operational tempo as is the case today, when experiential knowledge from the field is being generated that may conflict with previous doctrinal guidance. Moreover, in stability operations, junior officers experience the majority of interactions with local civilians and the enemy. One recent study by a war college identified dramatic on-the-job "discovery learning"[66] being accomplished by company commanders and other junior leaders in Iraq.

> The complexity, unpredictability, and ambiguity of postwar Iraq is producing a cohort of innovative, confident, and adaptable junior officers. Lieutenants and captains are learning to make decisions in chaotic conditions and to be mentally agile in executing counterinsurgency and nation-building operations simultaneously. As a result, the Army will soon have a cohort of company grade officers who are accustomed to operating independently, taking the initiative, and adapting to changes . . . The Army must now acknowledge and encourage this newly developed adaptability in our junior officers or risk stifling the innovation critically needed in the Army's future leaders.[67]

The capacity of the institution to tap into these lessons and translate them into institutional learning will demonstrate a cultural orientation toward the value of such bottom-up learning. To date, the only restrictions the military has placed on CompanyCommand.com is to insist that they use the domain extension *.mil* for their website and that the site be restricted to military personnel.[68] It is hard to argue with this decision,

---

65. Etienne Wenger, *Communities of Practice: Learning, Meaning, and Identity* (New York: Cambridge University Press, 1998).

66. The term *discovery learning* was used by junior officers of the 1st Brigade, 101st Airborne, Fort Campbell, KY, interviewed by the author, October 11–14, 2004.

67. Leonard Wong, *Developing Adaptive Leaders: The Crucible Experience of Operation Iraqi Freedom* (Carlisle, PA: U.S. Army War College, Strategic Studies Institute, July 2004).

68. According to the website, "CompanyCommand is a professional forum for U.S. Army Company-Level Commanders. . . . Membership is manually approved and is available only to Company-Level Commanders as well as currently commissioned officers who are either preparing for command or who have commanded in the past and desire to contribute to current company commanders." http://companycommand.army.mil/ev_en php?ID= 1_201&ID2=DO_ROOT.

given the ability of any potential adversary to log on and learn U.S. military tactics in an unrestricted forum.

Summary

This chapter clearly demonstrates that today's American Army is not the Army that fought in Vietnam. It is a high-tech, all-volunteer force whose learning processes have been dramatically transformed. Ironically, the post-Vietnam training systems designed to prepare the U.S. Army for a large land war against a Soviet enemy planted the seeds that are enabling the military to adapt for COIN and stability operations today. The leaders behind the scenes who designed and supported these systems did not all envision their use for anything other than "big war." Yet in the process, these leaders and the generation they led learned how to learn.

From a learning perspective, the development of the CTCs, along with the lessons-learned system it spawned in the 1980s, allowed the Army to "crack the code" on the problem of transforming a large, ponderous institution that had failed to adapt contemporaneously in Vietnam into a "learning organization" that responds rapidly to ongoing operations and facilitates the bottom-up transfer of information and dissemination of field knowledge. It is important to recognize that this training system did not occur by accident. Rather, it was the result of a concerted effort by a certain generation of leaders who consciously strove to create a learning organization. Indeed, these leaders studied learning theory and contracted the advice of organizational theorists from academia and the business world as they continued to try to improve upon the system.

Critics of today's operations in Iraq fail to recognize how much worse off troops in the field would be today had these systems and the lessons they generated throughout the 1990s not been in place. As discussed in chapter 6, from convoy operations to presence patrols, election monitoring, negotiation, refugee assistance, and interagency coordination, lessons learned from Somalia to the Balkans have been the "handrail" these leaders are using today. Moreover, the learning system described here allowed for the institutionalization of many of these lessons, despite ongoing political and strategic ambiguity over MOOTW and stability operations among and between military and civilian leadership throughout the 1990s.

Although it is still too soon to make a comprehensive analysis about ongoing operations in Iraq and Afghanistan, it appears that despite these obvious advances, there still is room for improvement in this system. Although the system has demonstrated great capacity for gathering bottom-up knowledge about tactics, techniques, and procedures, it is not clear that it has been as useful for the more complex, reconstruction-oriented operations American troops have been conducting in Iraq and Afghanistan. Rebuilding cities, kick-starting economies, and promoting democratic governance are not tasks that reveal immediate positive or negative feedback to troops on the ground that can then be fed back into a training environment. Seemingly "successful" reconstruction initiatives may demonstrate negative second-order effects at a later date. Thus, the current learning system may be inadequate for capturing these types of lessons. This book's conclusion reviews these issues, outlines successes and shortfalls in the current system, and provides recommendations for substantive changes and further study.

CHAPTER 5

Doctrine and Education for the New Force

Given the prevailing political climate of antipeace operations and the focus on "two major theater war" (2-MTW) strategy during the 1990s, one might not expect the military leadership to devote significant effort to developing doctrine and education for operations other than major war. Indeed, professional military education, as reflected by curricula at the midlevel Army and Marine Corps service schools, failed to make significant adjustments in response to the changes in the operational environment throughout the 1990s. This failure was not due to a lack of doctrine on the topic, however. In addition to older manuals that would have been relevant had they been resurrected, an entirely new crop of peace operations and "MOOTW" doctrine was developed. This chapter presents a detailed overview of the relevant MOOTW-oriented doctrine and the processes by which it was revised throughout the 1990s and beyond. In contrast to the post-Vietnam role of doctrine as an engine of military change, the process in the post–Cold War era, when the tempo of peace operations was high, reflects the role of doctrine as a trailing indicator of learning and an attempt to institutionalize lessons from the field. Thus, doctrine becomes both a trailing indicator of experiential adaptation and an engine of change.

The Changing Role of Doctrine and Education in the Military Learning Cycle

According to the Army's key field manual, FM 3-0 *Operations,* doctrine is the

> concise expression of how Army forces contribute to unified action in campaigns, major operations, battles, and engagements . . . It facilitates communications between Army personnel no matter where they

serve, establishes a shared professional culture and approach to operations, and serves as the basis for curriculum in the Army school system... To be useful, doctrine must be well known and commonly understood.[1]

In short, doctrine is what armies are supposed to know.

Formal published military doctrine is supposed to guide professional military education and is thus considered by military leaders as an "engine of change."[2] In theory, as new ideas or doctrinal theories are developed, recorded in manuals, and introduced into the education and training systems, actual military practice should change. Thus, military theorists who wish to change the way an army fights will rewrite the doctrine and see that the revised version is taught to the next generation. Indeed, as described in the previous chapter, this is exactly what happened after the Vietnam War when the AirLand Battle Doctrine was introduced. This doctrine, which proposed an integrated air and land offensive for fighting the Soviets, was a significant shift from previous doctrine and became the cornerstone for military education, training, and exercises for over a decade. The conventional military wisdom was that the AirLand Battle Doctrine was "validated" through experimentation and in the training centers throughout the 1980s and then in real-life operations during Operation Desert Storm in the 1990 Gulf War.[3] Subsequent 21st-century experience in Iraq and Afghanistan has called this into question and generated debate. Where the AirLand Battle Doctrine's enemy-focused concepts emphasizing speed and maneuver might have helped in winning the short-term military fights to topple the Taliban and the Iraqi military, they were insufficient or inappropriate in finishing the job and winning the peace in these population-focused conflicts.

The story of the progression from AirLand Battle Doctrine to Desert Storm presents the ideal model for doctrine development: a theory is proposed, debated, and recorded; it is tested and "validated" in experiments, practiced in exercises, and disseminated as doctrine throughout the system; finally, it is employed on the battlefield. This ideal model still

---

1. Department of the Army, *Field Manual 3-0 Operations (Formerly FM 100-5)* (Washington, DC, 2001).

2. Briefing by Lieutenant General David Petreaus, Commander, Army Combined Arms Center, Fort Leavenworth, Washington, DC, September 28, 2006.

3. Brigadier General David Fastabend, Deputy Director, Army TRADOC Futures Center, numerous personal interviews by author, 2004–6, Washington, DC; Colonel (USA Ret.) Robert Killebrew, personal interview by author, June 10, 2005, Basin Harbor, VT; James Kitfield, *Prodigal Soldiers: How the Generation of Officers Born of Vietnam Revolutionized the American Style of War* (New York: Brassey's, 1997); John L. Romjue, "The Evolution of the Airland Battle Concept," *Air University Review* (May–June 1984).

drives doctrine development today. A formal system for "future concept" (theory) development is run by the Joint Staff and Joint Forces Command (JFCOM) and is based on the scientific method. A joint operating concept (JOC) is written that theorizes a way of warfighting for a future environment (10 to 20 years). Once tested and validated through a rigorous series of experiments, the concept is used to inform the development of doctrine, education, training, and capabilities development (force structure).

Writing of the Army's attempt to remake itself at the end of the Cold War, General Gordon Sullivan claimed, "The process of writing new doctrine was the engine that drove the Army's transformation in its largest sense."[4] As with the Army's successful attempt to reorient itself to the post-Vietnam AirLand Battle Doctrine, the process by which Sullivan directed the revision of the Army's most important manual, FM 100-5 *Operations* (published in 1993), exemplifies how doctrine can be used by visionary leadership to institutionalize learning and drive change. Although the new manual did reflect a few of the lessons learned from Operation Just Cause in Panama and Operation Provide Comfort, the humanitarian food drop in Northern Iraq, the overall mission of Sullivan's effort was to prepare for the future. The entire four-star leadership of the Army met in a series of high-level conferences at which they debated the future of the Army and the proper focus for the new doctrine. They directed experiments—mostly at the higher headquarters level—from which they derived lessons about the future of warfare, and they developed a new vision for the service. Thus, the mission of the new doctrine was to guide the Army toward that new vision.[5] This is an example of top-down change, utilizing doctrine as a tool.

The problem with this ideal scientific method is that outside of peacetime, the process is too slow and tends to ignore the "real-life laboratory" of ongoing operations. This was a dilemma during the 1990s as doctrine and concept writers struggled to get ahead of events and incorporate the rapidly generated lessons from the field into this overly methodical process. Thus, Sullivan's capstone doctrine described above proved transitory; or in military terms, it was quickly "O.B.E." (overtaken by events). Throughout the 1990s, the leadership debated the nature of warfare and the role of the future military force, even as American troops con-

---

4. Gordon Sullivan and Michael V. Harper, *Hope Is Not a Method: What Business Leaders Can Learn from America's Army* (New York: Random House, 1996).

5. General (Ret.) George Joulwan, USA, personal interview by author, November 17, 2003, Arlington, VA; General Gordon Sullivan, USA, personal interview by author, February 25, 2005, Arlington, VA.

tinued to deploy at a constant pace—without visionary formal doctrine to guide their actions.[6] Sullivan's 1993 field manual underwent a number of controversial revisions before being officially revised and released in 2001.

More recent 21st-century doctrinal revisions, described below, reflect a concerted effort by the post-Vietnam generation of military leaders to get ahead of this cycle. Pushed by Iraq veterans such as Lieutenant General William Caldwell and General David Petraeus, the process of doctrinal development built on lessons from the 1990s and has been accelerated and actively crafted to incorporate lessons from the field. Moreover, the process of doctrinal dissemination was enhanced through the use of commercial presses, senior leader outreach, the use of the Internet, and active incorporation into educational as well as training curricula. General Petraeus described this cycle in his 2006 PowerPoint road show as an "engine of change." Both he and his successor, Lieutenant General Caldwell, actively drove this engine from their position as commander of the Combined Arms Center in Fort Leavenworth, Kansas.

Doctrine in this era has thus ceased to be a leading indicator, or "engine," of change based on theory but, instead, has become a trailing indicator of experiential adaptation and a tool to institutionalize these lessons from the field. Instead of being driven as a top-down process informed by theory, the critical doctrinal development of this era became more bottom-up, often driven by experienced midlevel officers. Nontraditional senior generals such as David Petraeus, William Caldwell, and James Mattis, who sponsored revisions of key manuals following their experiences in Iraq, have demonstrated how high-level leadership can play the key role in institutionalizing such bottom-up adaptation and learning.

Writing MOOTW and Peace Operations Doctrine in the 1990s

Despite ambiguity and debate at the strategic and political levels as well as the slow adaptation in the educational system, changes in formal MOOTW-oriented doctrine continued at a steady pace throughout the 1990s. These changes were reflected in the new and revised Army and joint field manuals described here. This parallel development of joint publications along with service-level manuals during the 1990s reflected the steadily growing momentum for improving joint operations,

---

6. Fastabend, interviews.

spawned by the Goldwater-Nichols Act in 1986, along with a struggle to understand the new post–Cold War operating environments.

Much of the joint and service doctrine published in the 1990s was derived from operational lessons learned in Panama, Somalia, Haiti, and the Balkans.[7] Indeed, references to specific events from these operations are a common thread throughout the publications. The pace and manner in which these manuals were developed and revised demonstrate how the doctrine-writing process, however slow compared to actual operations, effectively acted as a counterweight to education and strategic guidance that were even more out of touch with operational reality. Still, although the final products provided new guidance for conducting MOOTW in general and peace operations in particular, the language was careful to warn commanders not to let such missions inhibit unit readiness for "real" war.

Two competing forces acted on the doctrine-writing process in the 1990s. On the one hand were the strong strategic and political messages suggesting that these types of missions were a distraction for the U.S. military. Debates over readiness and politicized government studies demonstrating how peace operations eroded warfighting competence left many military leaders believing that MOOTW were a passing post–Cold War fad not requiring serious adjustments to strategy, structure, or doctrine.[8] Meanwhile, unit commanders tasked with training and deploying their units in real-world missions, such as in Haiti and the Balkans, needed practical guidance. As we will see here and in chapter 6, the system—especially when enhanced by visionary leadership—was designed to respond to their needs. When unencumbered by reactionary leadership, the system managed to work fairly well to produce relevant doctrine.

The system worked because this new generation of midlevel commanders, having been trained in the CTC system described in chapter 4, actively sought answers when training their units and, upon redeployment, shared their lessons learned. This phenomenon was aided at times by a handful of sympathetic senior leaders, such as Army chief of staff General Gordon Sullivan and Central Command (CENTCOM) commander General Anthony Zinni, who supported institutions and processes designed to generate knowledge and capture lessons. Still, there were other military and civilian leaders who were loath to see the

---

7. Donald G. Rose, "FM 3-0 Operations: The Effect of Humanitarian Operations on U.S. Army Doctrine," *Small Wars and Insurgencies* 13, no. 1 (2002).

8. Tammy Schultz, "Ten Years Each Week: The Warrior's Transformation to Win the Peace" (PhD diss., Georgetown University, 2005, AAT 3230095).

military reorient for what they considered inappropriate mission sets. The net result of these competing influences was a series of doctrinal publications that reflected operational realities but that also presented mixed messages for commanders tasked with peace operations.

The remainder of this chapter outlines the myriad substantive changes in formal published MOOTW doctrine since 1990. The processes by which many of these publications were written and published and the way in which they were interpreted reflect a military struggling to make sense of the post–Cold War world. Contrary to conventional wisdom, however, the fight over the development of peace operations doctrine and MOOTW missions did not fall neatly along civilian and military lines. While there were those in uniform as well as those in the civilian world who felt MOOTW was not a mission for military troops, others in uniform—often allied with civilian academics, pundits, and practitioners—continued to study MOOTW, write doctrine, and push for a more integrated understanding of the nature of military conflict. Current critiques notwithstanding, it is easy to see that without the efforts of this group of MOOTW and peace operations advocates toiling in the dark for so many years, commanders operating in Iraq and Afghanistan would have had even fewer resources to guide their actions.

Available MOOTW Doctrine at the End of the Cold War

When U.S. troops landed in Panama (for Operation Just Cause) in 1989, in northern Iraq (for Operation Provide Comfort) in 1991, and in Somalia (for Operation Provide Hope) in 1992, existing formal doctrine for the type of operations they were about to face was slim (see table 3). The latest version of the Army's primary manual for military operations, FM 100-5 *Operations,* had been updated as recently as 1986 but was focused on the offensively oriented AirLand Battle Doctrine and so contained no references to stabilization tasks. The Army's doctrine for military operations in urban terrain, FM 90-10, had not been updated since 1979 and was not widely known. Likewise, FM 100-20 *Military Operations in a Low-Intensity Conflict* was under revision at the time of the invasion of Panama by the new Army–Air Force Center for Low Intensity Conflict in Langley, Virginia. Although the new version would be released in December 1990, its distribution would have been limited at the time of the mission.[9]

---

9. Department of the Army and Department of the Air Force, *FM 100-20/AFP 3-20 Low Intensity Conflict* (Washington, DC, 1990); Guy Swan, "Swan on Swain," *Military Review* 5 (May 1988).

The previous edition of FM 100-20, published in 1981, had been modified in an interim "field circular," FC 100-20, published by the Army's Command and General Staff College in 1986, but distribution of the 100-20 series was still limited mostly to the small Special Operations community and still not relevant to the missions they were increasingly beginning to face.[10] Expressing his frustration at the inadequacy of available doctrine to the tasks at hand in the late 1980s, one Army officer wrote,

> "Everyone" has read and generally understands FM 100-5 Operations, but how many [non–Special Operations officers] will confess to even a rudimentary knowledge of FM 100-20? In reading both these manuals, one gets the impression that [low-intensity conflict] is merely AirLand Battle fought in a Third World Country.[11]

Meanwhile, although the Marine Corps had not been included in the newly formed congressionally directed Special Operations Command (SOCOM), the Corps had responded to the new interest in low-intensity conflict (LIC) by rereleasing its 1940s *Small Wars Manual* in 1987. Still, as stated in the document's foreword, the release was for "information only and [was] not directive in nature." In addition to the manual being quite dated, its dissemination was slim.[12] The primary focus of the Marine Corps at the time was on General Gray's new concept of "combined arms maneuver warfare," introduced in the 1989 Fleet Marine Force Manual 1 *Warfighting,* which, like the Army's AirLand Battle Doctrine, emphasized the offensive fight to defeat a military force with speed and combined arms.[13] Thus, for both services at the time, doctrine for what would become known as MOOTW, LIC, COIN, or Stability Ops was both quantitatively and qualitatively less than helpful.

TABLE 3. Available Relevant Doctrine, 1992

Manual	Year	Notes
FM 100-5 Operations	1986	Focused on Airland Battle Doctrine
FM 90-10 Urban Operations	1979	Not widely distributed
FM 100-20 Low Intensity Conflict	1981	Under revision; not widely distributed; mostly a Special Forces manual

---

10. Joseph Collins, former Deputy Assistant Secretary for Stability Operations, numerous personal interviews by author, 2004–5, Washington, DC; Wray Johnson, *Vietnam and American Doctrine for Small Wars* (Bangkok: White Lotus Press, 2001).
11. Swan, "Swan on Swain."
12. USMC, *Small Wars Manual of the United States Marine Corps* (1940; repr., Washington, DC: Government Printing Office, 1987).
13. USMC, *FMFM-1 Warfighting* (Washington, DC: Department of the Navy, 1989).

Adequacy of Available Doctrine

A perusal of the 1990 manual on LIC (the closest thing to MOOTW at the time) reveals that doctrine was still not quite adequate for the challenges faced in Panama, northern Iraq, or Somalia. LIC doctrine was broken into four categories: support for insurgency or counterinsurgency, combating terrorism, peacekeeping, and peacetime contingency operations. The 1990 version of FM 100-20 stated,

> When the United States uses military power directly against a hostile force in strikes or raids, the principles of combat operations govern tactical actions even though they occur in an environment short of declared war and are significantly influenced by constraints of policy and strategy.[14]

Thus, since Panama was to be a high-intensity, unilateral, combat-oriented invasion, the new doctrine for LIC would have seemed inappropriate for it anyway. Likewise, for humanitarian operations in northern Iraq and Somalia, there was little in FM 100-20 to direct complex civil-military efforts.

Numerous articles and after-action reports on Operation Just Cause noted the lack of appropriate doctrine for the challenges faced by U.S. troops in Panama.[15] Specifically, reports noted the need to develop and teach better doctrine for combat military operations in urban terrain, as well as for dealing with civilians, the media, nongovernmental organizations, and other elements not normally found on a traditional combat battlefield. Indeed, one after-action report published by RAND Corporation claimed that the success of Operation Just Cause was mostly due to highly favorable circumstances not likely to be encountered in subsequent operations.

> Had U.S. forces faced stiffer [Panamanian Defense Force] resistance in Panama City, for example, they would probably have found that they had received inadequate preparation and training for military operations in urban terrain (MOUT). Had U.S. forces encountered violent or even passive civilian opposition to the invasion, they could

---

14. Department of the Army and Department of the Air Force, *FM 100-20/AFP 3-20 Low Intensity Conflict*.

15. Ronald H. Cole, *Operation Just Cause: The Planning and Execution of Joint Operations in Panama* (Joint History Office, Office of the Chairman of the Joint Chiefs of Staff, 1995); Jennifer Morrison Taw, *Operation Just Cause: Lessons for Operations Other than War* (Santa Monica, CA: RAND Corporation, Arroyo Center, 1996).

have found themselves involved in an unconventional urban conflict requiring manpower they did not have and riot control, MOUT, and counterinsurgency operations for which they were neither trained nor prepared.[16]

In sum, for those who thought a highly trained combat force could conduct "smaller" operations without changing doctrine, education, or training, Panama should have been a wake-up call. Operation Provide Comfort and Operation Restore Hope only confirmed this reality in the three years that followed. Between the invasion of Panama and the 1993 publication of the revised FM 100-5 *Operations* and the brand-new JP (joint publication) 3-0 *Doctrine for Joint Operations*, U.S. troops faced similar challenges in northern Iraq and Somalia. Guided by the offensively oriented AirLand Battle Doctrine paradigm, it is no surprise that "the Army adopted an aggressive approach in carving out security zones in northern Iraq and establishing control of southern Somalia."[17] Likewise, a quick offensive to seize the port, the airfield, and the U.S. embassy characterized the Marine Corps' landing in Somalia, where a severe paucity of intelligence, cultural awareness, and strategic guidance frustrated the planning efforts by Marine Expeditionary Unit commander Colonel Greg Newbold.[18]

Upon landing and assessing the situation in Somalia, marines still found themselves without the doctrine, capacity, or authority to carry out missions such as food distribution and the capture and detention of warlords.[19] As a result of such frustrations, valuable lessons for dealing with NGOs, refugees, other civilians, and allies were learned in these two operations and would be reflected in subsequent MOOTW manuals developed throughout the decade. Unfortunately, some of the "lessons" recorded into doctrine were based more on a hopeful vision of future operations than on reality. Specifically, JP 3-08, which outlines interagency operations, suggests myriad tasks civilian agencies would be ex-

---

16. Taw, *Operation Just Cause*.
17. Rose, "FM 3-0 Operations," 71. For a discussion of the inappropriate use of force see Appendix C, "Operations in Somalia," in Department of the Army, *Field Manual 3-06 Urban Operations (Formerly FM 90-10)* (Washington, DC, June 1, 2003); C. Kenneth Allard, *Somalia Operations: Lessons Learned* (Washington, DC: National Defense University Press, 1995).
18. General (Ret.) Tom Clancy, USMC, Anthony Zinni, and Tony Koltz, *Battle Ready* (New York: Putnam, 2004); Rose, "FM 3-0 Operations." See also numerous news reports, December 1992, *Washington Post* and *New York Times*. Lieutenant General (Ret.) Gregory Newbold, USMC, numerous personal interviews by author, June–July 2005, Arlington, VA.
19. Newbold, interviews.

pected to perform in future Somalia-like operations. These ideas emerged through after-action seminars and simulations that sought to improve civil-military coordination for future operations. As discussed in chapter 6, the fact that such agencies lacked the capability and capacity to conduct these tasks was not clear to the theorists and doctrine writers at the time but would become painfully evident during the Iraq campaign starting in 2003. In 1993, however, compared to previous eras, it was significant that the new FM 100-5 and JP 3-0 mentioned these types of missions and these myriad nonmilitary actors at all.

Doctrine Development for the New World Order

The 1993 versions of JP 3-0 and FM 100-5 were the first post-Vietnam operations manuals to include entire chapters on "operations other than war." These mark the beginning of a new generation of manuals that sought to capture lessons from ongoing operations of the 1990s (table 4). Previous versions of FM 100-5 in the 1960s had contained chapters on "unconventional warfare operations," "military operations against irregular forces," and "situations short of war" (as part of President

TABLE 4. The Next Generation of MOOTW and Stability Operations Doctrine

Manual	Year	Notes
FM 100-5 Operations	1993	Revised to include MOOTW chapter
JP 3-0 Joint Operations	1993	Included MOOTW chapter
FM 100-23 Peace Operations	1994	Promoted by Army's Peacekeeping Institute
JP 3-07 MOOTW	1995	Entire joint manual devoted to MOOTW
JP 3-08 Interagency (vols. 1 and 2)	1996	Incorporated lessons from previous operations
Joint Task Force Commander's Handbook for Peace Operations	1997	Operational level manual for commanders
JP 3-07.3 Tactics, Techniques, and Procedures for Peace Operations	1999	Tactics, techniques, and procedures for peace operations
FM 3-0 Operations	2001	Introduced "Full Spectrum Operations"; replaced *FM 100-5*
FM 3-07.31 Tactics, Techniques, and Procedures for Peace Operations	2003	Multiservice manual developed by ALSA with Peacekeeping Institute
FM 3-07 Stability and Support Operations	2003	Replaced the *MOOTW* term
FM 3-24 Counterinsurgency	2006	Driven by General Petraeus
FM 3-0 Operations	2008	Socialized "Full Spectrum Ops"
FM 3-07 Stability and Support Operations	2008	Published by University of Michigan Press

Kennedy's attempted emphasis on counterinsurgency) but still were offensively focused.[20] No mention of anything other than major warfare had been included in this critical manual since the end of the Vietnam War. In 1982, AirLand Battle Doctrine was introduced and remained the focus through the 1986 revision.[21] The following discussion outlines the substantive changes made in some of the key manuals listed in table 4 and the behind-the-scenes debates and processes by which they were published.

Army FM 100-5 *Operations*

In 1993, FM 100-5 acknowledged, "Today, the Army is often required, in its role as a strategic force, to protect and further the interests of the United States at home and abroad in a variety of ways other than war." Chapter 13, "Operations other than War," signaled an expansion from the concepts developed in LIC doctrine. Perhaps more important, by including this chapter in FM 100-5, the Army's primary operations field manual, the authors were acknowledging that MOOTW/LIC was no longer a mission for the Special Operations Forces alone. Yet with the opening sentence of the chapter stating clearly that "the Army's primary focus is to fight and win the nation's wars," the strategic ambiguity and political debate over using U.S. forces for MOOTW rang loud and clear.

Far from being the "bottom-up" organizational change advocated by some learning theorists, FM 100-5 was a deliberate attempt by the senior Army leadership to drive the organization to adapt to the changing world order. According to General Gordon Sullivan, chief of staff of the Army at the time, this revision corresponded to a critical time in Army history that required senior leadership to "drive change."[22] The combination of the end of the Cold War, dramatic downsizing, and an increased operational tempo requiring back-to-back deployments made it clear to Sullivan that doctrine needed to be updated. Moreover, the difficulties the Army had had in mobilizing for Desert Storm convinced him that the new doctrine had to work for a new, more mobile, rapidly deployable Army. As discussed above, the result was a meeting of the se-

---

20. *FM 100-5* had been updated in 1968, 1976, 1982, and 1986. The 1982 version had introduced AirLand Battle Doctrine.
21. Department of the Army, *Field Manual 100-5 Operations* (Washington, DC, 1986); Department of the Army, *Field Manual 100-5 Operations* (Washington, DC, 1982).
22. Sullivan and Harper, *Hope Is Not a Method*.

nior Army generals in September 1991 in which they considered "whether it was time to rewrite Army doctrine for contingency operations, peacekeeping, power projection, and other aspects of the Army's rapidly evolving missions."[23] This group of leaders was attempting to do for their generation what AirLand Battle Doctrine did for the generation before—lead change.

According to accounts and interviews of some of the key players, the addition of the MOOTW chapter was not a unanimous decision among this group of the Army's most senior generals. As Sullivan explained, there were those in the room who felt that "we had just demonstrated [in the Gulf War] that we were the world's best Army, so if we needed to change and mix up the artillery or something, fine, but nothing more."[24] Other key players in addition to Sullivan—including General George Joulwan, commander of Southern Command; General Carl Steiner, commander of Special Forces Command; and General Fred Franks, commander of Training and Doctrine Command—became the strongest and most influential advocates of including a chapter on MOOTW.[25] While General Sullivan's approach was not to order the doctrine writers to include the chapter, he admits that it was "no mystery" what his wishes were. General Joulwan, however, took a more activist approach.

As the commanding general of Southern Command (SOUTHCOM) at the time, Joulwan was charged with leading the "drug war" in the Western Hemisphere. Recognizing that NGOs had expertise and assets that the military did not possess, Joulwan learned quickly that success in such nonstandard environments required the coordinated efforts of these disparate entities. Because he had come to believe very strongly that future U.S. military operations were more likely to resemble his multinational, interagency, SOUTHCOM experience than the force-on-force efforts in the Gulf War, he advocated that the MOOTW chapter be included as the opening chapter to the new FM 100-5. In the end, the MOOTW chapter was included as chapter 13. Not surprisingly, many of the lessons learned from Joulwan's experience in the drug war (i.e., on interagency cooperation, the importance of understanding NGOs, etc.) were directly reflected in the MOOTW chapter.[26]

---

23. Sullivan, interview; Sullivan and Harper, *Hope Is Not a Method*.
24. Sullivan, interview.
25. Richard Duncan Downie, *Learning from Conflict: The U.S. Military in Vietnam, El Salvador, and the Drug War* (Westport, CT: Praeger, 1998); Joulwan, interview; Sullivan and Harper, *Hope Is Not a Method*.
26. Department of the Army, *Field Manual 100-5 Operations* (Washington, DC, 1993).

Army FM 100-23 *Peace Operations*

The following year, the Army published a new, more specific, MOOTW manual, FM 100-23 *Peace Operations*. This publication directly cited President Clinton's Presidential Decision Direction 25 as justification for focusing specifically on peace operations.[27] Following FM 100-5's chapter on MOOTW, in which peace operations were discussed as a subcategory to operations other than war, FM 100-23 sought to provide more detailed doctrinal guidance for the "full range of peace operations, to include support to diplomacy (peacemaking, peace building, and preventive diplomacy), peacekeeping (PK), and peace enforcement (PE)."[28] Importantly, the manual "incorporates lessons learned from recent peace operations and existing doctrine."[29] Operations Provide Comfort and Restore Hope are referenced directly throughout the manual, indicating an attempt to institutionalize critical lessons from these experiences. Prior to FM 100-23, officers had been referencing an infantry school white paper, *The Application of Peace Enforcement (PE) at the Brigade and Battalion Level*.[30] Accordingly, the white paper informed the writing of FM 100-23.

FM 100-23 emphasizes planning and logistics. Appendixes are included to introduce officers to other nonmilitary actors likely to be present in the peace operations environment, such as the United Nations, U.S. government agencies, intergovernmental organizations, and NGOs. Although this manual provided a much-needed top-level resource for commanders, it was still quite thin on specific guidance. This lack of specificity extended to the guidance for peace operations training, where the manual conveyed more mixed messages to commanders charged with preparing their units for such missions.

In appendix C, "Training," FM 100-23 begins with the following familiar statement: "Training and preparation for peace operations should not detract from a unit's primary mission of training soldiers to fight and win in combat." It goes on to point out, "Peace operations . . . should not be treated as a separate task to be added to a unit's mission-essential task list (METL). However, units selected for these duties require time to

---

27. Department of the Army, *Field Manual 100-23 Peace Operations* (Washington, DC, 1994).
28. Ibid.
29. Ibid.
30. U.S. Army Infantry School Commandant, *White Paper: The Application of Peace Enforcement (PE) at the Brigade and Battalion Level* (Fort Benning, GA, August 31, 1993).

train and prepare for a significant number of tasks that may be different from their wartime METL."[31] Finally, as if to highlight the collective schizophrenia over how and if U.S. forces should prepare for these missions, the paragraph ends with this statement: "The philosophy used to determine the how much and when training questions for operations other than war can be summed up as *just enough* and *just in time*."[32] This confusing message suggests that the manual's authors were well aware of the political debates over MOOTW missions at the time. To suggest significant changes to unit METLs or training regimes would have been a highly controversial step. In the end, such decisions were left to individual commanders.

A few division commanders (i.e., the 82nd Airborne and the 25th Light Infantry) did take the initiative to add peace operations tasks to their training exercises. This move was not widely accepted throughout the military, however, and was eventually thwarted by a directive from the Forces Command commander in 1999 stipulating that division commanders would not be allowed to add noncombat tasks to their METLS unless and until their unit was specifically scheduled to deploy for such a mission. This directive was reflected in the 2001 update to FM 100-5 (by then renumbered to FM 3-0), which reinforced the ambiguity over doctrine, training, and operations.

JP 3-07 *Military Operations other than War*

By the mid-1990s, in addition to the new chapters on MOOTW in FM 100-5 and JP 3-0 and a brand-new manual for peace operations, FM 100-23, Army and Marine Corps commanders deploying to Haiti in September 1995 and to the Balkans in December of that same year should have had access to a new joint-level manual focused specifically on MOOTW, JP 3-07 *Military Operations other than War*, published earlier that summer. This manual built on the MOOTW concept and provided an operational-level overview of MOOTW.[33] In the manual's foreword, chairman

---

31. METL refers to the "mission essential task list" that each unit derives and maintains. The list identifies the tasks at which the unit must be proficient. See chapter 7 for more information on METLs.

32. Department of the Army, *Field Manual 100-23 Peace Operations*, 86.

33. U.S. Army, *FM 3-07 (FM-100-20) Stability and Support Operations* (Washington, DC: Headquarters, U.S. Army, February 20, 2003); Steve Capps, editor of the *FM 3-0* 2001 and 2005 editions, personal interview by author, February 8, 2005, via telephone and follow-up e-mails to Combined Arms Doctrine Division, Fort Leavenworth, KS; Mike Chura, doctrine writer and editor of *FM 3.07 Stability and Support Operations, Combined Arms Doctrine Division*, personal interviews by author, February 10, 2005, via telephone and e-mail, Orlando, FL.

of the Joint Chiefs of Staff General Shalikashvili stated, "While we have historically focused on warfighting, our military profession is increasingly changing its focus to a complex array of military operations—other than war . . . Participation in MOOTW is critical in the changing international security environment."[34] General "Shali" was well known for having said "real men don't do MOOTW," but by the mid-1990s, as troops were simultaneously preparing for Operation Uphold Democracy in Haiti and for IFOR in Bosnia, he seemed to have recognized that MOOTW were missions the military could no longer avoid.[35]

The audiences for JP 3-07 were senior officers and staff who would most likely be involved in planning and leading MOOTW missions. The manual provided a general overview of MOOTW to include an explanation of how MOOTW differed from traditional warfare, the various types of MOOTW officers might expect to conduct, and the myriad nonmilitary actors officers might encounter during these operations, such as NGOs, the State Department, and UN officials. A chapter on planning attempted to pull all of this together at the operational level by suggesting ways to integrate these actors into a comprehensive operational plan. Unfortunately, these other actors were not part of the doctrine-writing process, which resulted in inaccurate assumptions being made in their absence about what they would actually be capable of doing should the time come.

Again, mixed messages regarding MOOTW and warfighting were evident in JP 3-07, especially in the slim, four-paragraph discussion of education and training. The section claims, "For some MOOTW (for example, humanitarian assistance and peacekeeping operations) warfighting skills are not always appropriate." Yet the manual notes that training for these nonwar skills will be difficult because of the military's primary role to "fight and win the nation's wars." The solution offered for this conundrum is that education of officers and NCOs on MOOTW principles and types must make up for this training shortfall: "The lack of opportunity to train for a specific operation is in large part overcome by military leaders who have a solid foundation of MOOTW provided through the military education system." Thus, the military acknowledged that the majority of troops will likely learn these skills on the job—under the direction of well-educated officers and NCOs. Unfortunately, the education of

---

34. Department of Defense, *Joint Publication 3-07: Joint Doctrine for Operations Other than War* (Washington, DC: U.S. Government Printing Office, June 16, 1995).

35. Although General Shalikashvili had commanded Operation Provide Comfort, he was still widely known to be averse to MOOTW. General (Ret.) Anthony Zinni, USMC, personal interview by author, January 31, 2005, Arlington, VA.

officers during this period, which is discussed in more detail below, focused very little of its curriculum on the challenges of MOOTW.

Despite the familiar mixed messages, the publication of the joint manual for MOOTW was a significant development. By building on the six common principles of the many types of MOOTW, including "objective," "unity of effort," "security," "restraint," "perseverance," and "legitimacy," the publication of JP 3-07 laid the foundation for the development of future manuals, such as the two TTP ("tactics, techniques, and procedures") manuals—JP 3-07.3 *Joint Tactics, Techniques, and Procedures for Peace Operations* (1999) and FM 3-07.31 *Multi-service Manual TTP for Peace Operations* (2003)—and JP 3-08 *Interagency Coordination for Joint Operations*.[36]

These follow-on manuals continued to build on the foundations developed in the three capstone documents discussed above (FM 100-5, FM 100-23, and JP 3-07). They provided more detailed guidance for lower levels of command for specific peace operations tasks such as patrolling, checkpoints, convoy operations, and refugee management, to name only a few. The *Multi-service Manual TTP for Peace Operations* (2003) contained entire chapters on force protection, civil-military relations, and conflict resolution, as well as detailed appendixes on how to set up town meetings; how to work with liaisons, interpreters, the media, and NGOs; and much more. In short, it is clear that the multiservice manual was responding directly to operational realities at a more practical level as it documented ongoing lessons learned from the field.

"Full-Spectrum Operations": The 2001 FM 3-0 *Operations*

The 2001 Army operations manual, FM 3-0, replaced the 1993 FM 100-5 and reflected myriad lessons learned from the 1990s. General Shinseki claimed in the manual's foreword, "This edition has been shaped by our experiences and experiments since the first post–Cold War FM 100-5 published in 1993 and the duties we foresee for our Nation in this early

---

36. Air Land Sea Application Center, *Field Manual 3-07.31/MCWP 3-33.8/AFTTP(I) 3-2.40: Multi-Service Tactics, Techniques, and Procedures for Conducting Peace Operations* (Langley Air Force Base, VA, October 2003); Department of Defense, *Joint Publication 3-07.3: Joint Tactics, Techniques, and Procedures for Peace Operations* (Washington, DC: U.S. Government Printing Office, February 12, 1999); Department of Defense, *Joint Publication 3-08: Joint Doctrine for Interagency Coordination*, vol. 1 (Washington, DC: U.S. Government Printing Office, October 9, 1996); Department of Defense, *Joint Publication 3-08: Joint Doctrine for Interagency Coordination*, vol. 2 (Washington, DC: U.S. Government Printing Office, October 9, 1996).

part of the 21st century."³⁷ Thus the manual was intended not only to capture lessons from Somalia to the Balkans but to anticipate the conflicts of the future and an American way of war that was more "joint," more integrated with respect to other instruments of national power, and less predictable.

The new numbering system mirrored that introduced by the joint doctrine system, symbolically and conceptually bringing the next generation of Army publications in line with the joint community.³⁸ More important, however, the 2001 FM 3-0 introduced the concept of *full-spectrum operations,* which added "stability" and "support" to the familiar "offense" and "defense" elements of warfighting. The manual de-emphasized the entire concept of MOOTW, mentioning it only six times and never actually defining it. In fact, when the Army's writing team began its work in 1996, it was specifically directed by their immediate commanders not to use the term *MOOTW* in the new manual.³⁹ However, because MOOTW was an official term for joint doctrine, the writers had to adapt. One of FM 3-0's authors claimed, "We really didn't want to use [*MOOTW*], but because it was in joint doctrine, we acknowledged it."⁴⁰

The demonstrated distaste for the MOOTW term was born from recent experience that caused these doctrine writers and Army leaders to rethink their capstone doctrine. According to one of the manual's authors, the decision to switch to a full-spectrum approach "came right out of Bosnia," where experience indicated that the perceived separation between "war" and "peace" was not as clear for troops on the ground as it might be for strategists, politicians, and academics.⁴¹ Thus the downplaying of MOOTW reflected important lessons learned from new experience, as the authors intended to switch the focus to "describing what Army forces do versus the environment in which they operate."⁴² A "full spectrum" of operations that applied to various types of conflict—war or peace—would be the conceptual hook.

Lieutenant Colonel David Fastabend (a two-star general at the Army's

---

37. Department of the Army, *Field Manual 3-0 Operations (Formerly FM 100-5)*.
38. Colonel (Ret.) Michael D. Burke, USA, "FM 3-0: Doctrine for a Transforming Force," *Military Review* (2002); Michael Burke, member, writing team *FM 100-5*, 1996–2001, *FM 3-0* 2001 and 2005 editions, personal interview by author, February 3, 2005, via telephone to Combined Arms Doctrine Division, Fort Leavenworth, KS.
39. Burke, interview; Killebrew, interview.
40. Capps, interview.
41. Brigadier General David Fastabend, Deputy Director, Army TRADOC Futures Center, personal interview by author, February 8, 2005, Fort Monroe, VA.
42. Quotation from Capps, interview; Fastabend, interview.

Futures Center by 2005 and then a lead strategist in Iraq in 2006–7) was a key member of the writing team for FM 3-0 in 1996. That the full-spectrum concept bore a remarkable resemblance to ideas he had penned in an 85-page paper written as an Army Fellow at Stanford University's Hoover Institution in 1996 was no coincidence. Writing of his experience as a member of a lessons-learned collection team to Bosnia, the lieutenant colonel–scholar noted that U.S. peacekeeping forces there had been required to shift rapidly from high- to low-intensity operations. Clashing with war criminals, he observed, meant forces needed both the "invocation and demonstration of firepower," while at other times, simply patrolling or manning checkpoints was all that was required.[43] Unfortunately, Fastabend noted, doctrine did not adequately reflect this reality.

> Army doctrine, for example, currently addresses conventional operations in Field Manual (FM) 100-5—with a focus on the violence of heavy combat—and "operations other than war" with a focus on the logic of low-intensity conflict and peace operations in FM 100-20 and FM 100-23, respectively. The issue for services will be whether to continue to pursue this doctrinal "dual track," or to develop a single, overarching doctrine for both war and "operations other than war." A revolution in military doctrine would be one that bridges this dual track into a single unified approach that effectively encompasses the interdependent mechanisms of both *violence* and *logic* in conflict.[44]

Thus, full-spectrum theory was intended to rectify the conceptual "trap" that doctrine writers and military leaders felt the artificially dichotomous "war-MOOTW" distinction created. It may be the case that these two types of operations differ, legally, politically, and otherwise; but for troops on the ground, experience from Somalia to Bosnia had demonstrated that none of those distinctions mattered. Troops sent to save lives must be ready to defend their own; and peacekeepers that are not prepared or permitted to use force will eventually find themselves witness to or victims of violence.

Readers may recognize the similarity of this concept to Marine Corps General Charles Krulak's concept of "three-block war." Krulak suggests

---

43. Lieutenant Colonel David Fastabend, USA, *A General Theory of Conflict: Bosnia, Strategy, and the Future* (Hoover Institution, Strategy Research Project, Stanford University, 1996).
44. Ibid., 55.

that "marines may be confronted by the entire spectrum of tactical challenges in the span of a few hours and within the space of three contiguous city blocks," including "humanitarian assistance, peacekeeping, and traditional war-fighting."[45] Indeed, the similarity between full-spectrum and three-block war did not escape General Fastabend. "We were not as pithy as the 'three-block war,'" Fastabend said, "We just wrote it into doctrine."[46]

In sum, full-spectrum operations doctrine was meant to reflect the realities of soldiers on the ground. Importantly, it sought to identify the conceptual link between combat-trained forces and peace operations missions. Because these soldiers writing doctrine felt very strongly that "credibility in peacekeeping operations stems first and foremost from the potential enemy's certain conviction that the U.S. Army would defeat them if the situation arises,"[47] the Army would no longer think of MOOTW as something so inherently "other" than war. Conceptually, full-spectrum theory meant that all missions would require a combination of offense, defense, stability, and support operations. The commander on the ground was required to determine the appropriate proportional mix.

With this new doctrine, the Army as a whole would need to be organized, trained, and equipped to conduct more than just offensive combat maneuvers. Doctrinally, this meant that the next level of doctrine, FM 3-90 *Tactics*, which provided more detailed guidance on particular maneuvers referenced in the capstone FM 3-0, would need to be updated for stability and support operations. For a small number of officers who had spent the decade in Somalia, Haiti, and the Balkans and who had been "crying in the dark" about the need to elevate the status of peace operations doctrine, FM 3-0 and the full-spectrum theory created the conceptual bridge—and the political window—they had been waiting for to develop a more specific manual, Army FM 3-07 *Stability and Support Operations (SASO)*. Colonel George Oliver, director of the Army's Peacekeeping Institute and key promoter of the new manual, explained to me, "If it hadn't been for 3-0, nothing would have happened . . . We would not be anywhere today without [FM 3-0]."

---

45. General Charles C. Krulak, USMC, "The Strategic Corporal: Leadership in the Three Block War," *Marines Magazine* (January 1999).
46. Fastabend, interview.
47. Burke, "FM 3-0: Doctrine for a Transforming Force."

A Hothouse for Military Change: PKI and the Next Generation of Doctrine

The development of Army FM 3-07 *Stability and Support Operations (SASO)* and the next generation of peace operations doctrine was driven by a small group of midlevel officers at the Army's Peacekeeping Institute (PKI) in Carlisle, Pennsylvania. The story of PKI and its role in developing doctrine and pushing for peace operations in military education reflects the sustained post–Cold War tension over MOOTW and the way in which midlevel officers can influence the trajectory of military change. Like the marines in the Banana Wars who fought their service's preference for amphibious warfare and wrote the *Small Wars Manual* while teaching at Quantico, a similar group of experienced officers in the 1990s sought to record their lessons from peace operations into core Army doctrine. This effort was enhanced by the creation of a new Army institute focused on peacekeeping, which provided the organizational space for this intellectual work.

PKI was opened in 1994 at the direction of the chief of staff of the Army, General Gordon Sullivan. As Sullivan explains, the idea of the institute "was born out of Somalia," where the leaders on the ground were overwhelmed by the challenges of working with a proliferating body of NGOs. After visiting his officers there, many of whom told him outright, "Working with NGOs is hard," Sullivan was convinced that the Army as an institution had a lot to learn about such operations. Moreover, he thought the Army was being misunderstood and prejudged by nonmilitary actors who saw soldiers as one-dimensional "killers on the battlefield." He knew that to work more effectively in this multiagency environment, the Army not only would have to learn more about NGOs and peace operations in general but would also need to educate others about the "rich traditions of the Army." Thus the motivation for this institute was twofold: to generate knowledge and to improve the Army's image.[48]

Although others in the military saw Somalia as a sui generis event, General Sullivan felt that "this would not be a passing moment." No matter how hard the military leadership tried to will these missions away, chances were it would continue to be sent abroad into other-than-major-war scenarios. Moreover, ignoring the realities of such a volatile and unpredictable post–Cold War world would only foment future crises. According to General Sullivan, "saying, 'You can't do peacekeeping,' is like

---

48. Sullivan, interview.

saying, 'I'm going to leave all these petri dishes with this liquid in them and let viruses breed.'" The chief predicted that because MOOTW was the type of "task that the Army would perform well into this century," soldiers would eventually come looking for answers.[49]

Although PKI existed on the margins of the mainstream Army and had fought for its bureaucratic existence and funding year after year, the institute's founder, General Sullivan, knew that however small it was as an institution, it was an important first step. The establishment of the institute reflects the general's visionary leadership and, perhaps more important, his understanding of bureaucratic politics and Army culture. "Soldiers only eat when they are hungry," Sullivan noted, "and when they are hungry, they eat everything in sight. The Peacekeeping Institute would be a place they could go to be fed."[50] Noting that the "intellectual leads the physical," Sullivan's goal was to plant an intellectual seed that could germinate in the dark, even as much of the military leadership remained uninterested. The general knew that eventually the Army would come looking for answers to the peacekeeping puzzle. To the chagrin and dismay of many, and despite powerful attempts to close the institution, this is eventually what happened.

Not everyone in positions of authority shared General Sullivan's vision. With the election of the Bush administration in 2000 came increased scrutiny on the activities of PKI. Their funding was small to begin with (less than one million dollars annually),[51] and their existence had never exactly been a priority, even in the Clinton administration,[52] but in the new environment, PKI found itself answering inquiries for myriad Pentagon "studies" designed to ascertain exactly what the institute was doing and how much it was costing the Army. By 2003, in accordance with the new Bush administration's avowed distaste for peacekeeping, the small institute knew it was "on the chopping block."[53] Finally, in January 2003, to no one's surprise, the Army announced that PKI would close its doors that October.

The official story published in the press was that it was a cost-saving move and that because the PKI functions were redundant, they could be easily absorbed by the existing Center for Army Lessons Learned

---

49. Ibid.
50. Ibid.
51. Foster, "Pentagon Peacekeeping 101," *Milwaukee Journal Sentinel*, July 14, 2003.
52. Colonel George Oliver, U.S. Army, Former Director, Army Peacekeeping Institute, numerous personal interviews by author, October 2003–June 2005, Washington, DC.
53. Lorelei Kelly, Stimson Center, "A Military Orphan Faces the Ax," *Boston Globe*, April 26, 2003.

(CALL) in Fort Leavenworth, Kansas. In reality, a few boxes from PKI had been sent to CALL after the announcement was made, but nothing was being done with them, and nothing other than filing the documents was planned. CALL had a tactical focus, not a strategic or interagency one. Officials there had their hands full as it was, simply keeping up with ongoing operations and capturing lessons in the field. The Army had given CALL no additional resources or manning to accommodate the added PKI functions and thus, to all interested observers, had clearly intended to kill the PKI program.[54] Only vigorous lobbying and behind-the-scenes activities by a coalition of think tank experts, members of Congress, and a few uniformed officers, who together pointed out the need for the institute in light of the unfolding debacle of postwar Iraq in the summer of 2003, saved the institute from this planned death.[55]

In 2004, Colonel John Agoglia, a talented war planner with a reputation for being a maverick, took over as director of the newly resuscitated institute. Reflecting the influence of current operations as well as new doctrine and the intent to expand the focus of the institute, Colonel Agoglia renamed it the "Peacekeeping and Stability Operations Institute" (PKSOI) and set out to bring the organization back to life. Agoglia's mandate was to create a "Center of Excellence" for peace and stability operations issues, which meant reaching out to other agencies and think tanks working in the field. With his tireless enthusiasm and talent for meeting people and connecting them, Agoglia was an excellent candidate for this task. Within two years, PKSOI had become a focal point for interagency coordination for stability operations and a key coordinator of stability operations doctrine, education, and training.

In 2005, when the deputy assistant secretary of defense signed a new DoD directive (DoDD 3000.05)[56] mandating that stability operations be given priority comparable to combat operations, the Army authorized a significant expansion of PKSOI. In this new political climate, Agoglia grew the fledgling six-person, $750,000 Army think tank into a well-respected institute with 49 officers, civilian practitioners, and academics and an annual operating budget of three million dollars. As predicted by Sullivan in the mid-1990s, PKSOI had effectively kept the candle burn-

---

54. Site visits to CALL and PKI and numerous interviews with CALL officials, March 2004, and PKI personnel, January 2004; Kelly, "A Military Orphan Faces the Ax."

55. Jack Kelly, "Iraq Provides Peacekeeping Institute with Needed Boost," *Pittsburgh Post-Gazette*, November 27, 2003.

56. Deputy Secretary of Defense Gordon England, *DoD Directive 3000.05: Military Support to Stability, Security, Transition, and Reconstruction Operations (Sstro)* (Washington, DC, 2005).

ing on the topic for a decade, and now the institute's knowledge and resources were actively being sought.

The Peacekeeping Institute and Doctrine Development

PKI's role in doctrine writing began in the mid-1990s. Colonel George Oliver, the third director of the Army's Peacekeeping Institute, together with his chief doctrine writer, retired Army officer Bill Flavin, influenced the development of a number of new doctrinal publications throughout the 1990s. Their focus on doctrine development stemmed from their understanding of the links between doctrine, education, and military change. Oliver explained, "We knew we had to get the doctrine straight in order to get it into the education system."[57]

The PKI team actively sought to fill the gaps in MOOTW doctrine by assisting in the long development process of a few key publications. They infiltrated the doctrine-writing processes and catalyzed the development of new manuals by hosting a series of meetings to which they invited interested parties from the Marine Corps, the Army infantry school, TRADOC, and the Air Force. Once writing began on key publications such as the keystone Army manual *Stability Operations* (FM 3-07) and *Multiservice TTP for Peace Operations* (FM 3-07.31), the institute—and especially Bill Flavin—provided a great deal of the conceptual thinking, drafted outlines, coordinated inputs from interested participants, and "fed whole chapters to Fort Leavenworth," where they were scrutinized and edited by the writing team at the Center for Army Doctrine Development. The final products filled a critical doctrinal gap—one that the experts at the Army's Peacekeeping Institute were uniquely suited to address. PKSOI would continue to facilitate subsequent revisions of these manuals into the next decade.

Starting in the mid-1990s and continuing to today, PKSOI's team of officers and scholars have acted as expert interlopers in the doctrine-writing system. Thus, they were able to catalyze and contribute to the development of new doctrine. From the perspective of organizational learning theory, this example highlights the importance of institutional structure. Creating a "hothouse" in which fledgling ideas could germinate and "thought leadership" could take place was a critical step in this process. From the perspective of military innovation, this example reveals the influence of bureaucratic politics and how savvy, visionary political actors (i.e., General Sullivan as PKI's founder and Colonel

---

57. George Oliver, personal interviews by author, June 15, 2005, Washington, DC.

Agoglia) can set a trajectory for change through small structural or institutional initiatives that influence the learning system of the organization.

FM 3-24 *Counterinsurgency*, FM 3-07 *Stability Operations*, and the New Era of Military Doctrine

The process of doctrinal development shifted significantly in response to operations in Iraq and Afghanistan. Strong senior leadership from well-placed advocates such as General Petraeus, Lieutenant General Caldwell, and Lieutenant General Mattis provided the extra leverage mid-level doctrine writers needed to mainstream their ideas into the institution. The development of two key Army field manuals, FM 3-24 *Counterinsurgency* and FM 3-07 *Stability Operations*, reflects this fundamental shift.

The writing of the 2006 COIN manual, FM 3-24, began with a call for help from the field. When CENTCOM commander General John Abizaid relieved General Tommy Franks, he immediately requested existing counterinsurgency doctrine from the Special Operations school in Fort Bragg, North Carolina. The request launched an accelerated doctrine-writing process to produce an "interim" field manual—FMI 3-07.22 *Counterinsurgency Operations*—that could be distributed widely for comment and immediate use via the Internet. The interim manual was written in an unprecedented six months through a series of workshops by a group of subject matter experts from the Special Operations community, academics, and veterans, all corralled by Major Jan Horvath, serving under General Wallace, commander of the Combined Arms Center at Fort Leavenworth. This episode reflected the "big Army's" attempt to harvest the seeds that had been germinating inside the Special Operations community since the late 1980s and its quest to get something to the field as quickly as possible.

The interim manual was widely circulated—and critiqued—in the months that followed. The next phase sought to incorporate these critiques and to bring other services and nonmilitary government agencies into the process, to turn the "interim" field manual into a proper interservice field manual. This effort was led by Lieutenant General Petraeus beginning in the fall of 2005. Petraeus appointed Dr. Conrad Crane, an Army historian, former West Point classmate, and counterinsurgency expert, as the primary editor. Crane worked with Lieutenant Colonel John Nagl to outline the manual and identify other experts who could write or vet various chapters. The draft manual was vetted via a series of workshops and conferences with a large community of military and nonmilitary academics, journalists, and practitioners, including civil servants, aid workers, development NGOs, and human rights activists. "Bootleg" copies were disseminated widely beginning in the spring of 2006, and the manual was released later that fall. Finally, the Army worked with the University of Chicago Press to publish the manual and make it available to an even wider audience.

FM 3-07 *Stability Operations* was published in the fall of 2007 and followed a similar process. The writers sought to avoid the trap discussed above—of writing about other departments and agencies and assuming responsibilities and tasks for them without actually consulting with them. Lieutenant Colonel Steve Leonard, the chief writer of the document, actively sought the perspective of other agencies as well as nongovernmental organizations. He wanted to know what their actual capabilities were as well as how they viewed themselves operating as part of the comprehensive interagency approach in the field. Through a series of conferences, roundtables, and workshops with thought leaders and representatives from various agencies throughout the government, in the NGO community, and among allies, Lieutenant Colonel Leonard was able to glean the latest thought, theory, lessons, and controversies from the widest possible group of experts. Detailed debates over language, connotations, social science theory, and recent lessons learned from the field took place over a 10-month period, with some of these nonmilitary participants contributing actual text to the finished product. The goal of Lieutenant Colonel Leonard's commander, Lieutenant General Bill Caldwell, was that the manual would be adopted by these other organizations and agencies as their own, even though the process was sponsored by the U.S. Army. Like FM 3-24, the stability operations manual, FM 3-07, was published by a major university press for greater distribution beyond the U.S. Army.[58]

Compared to the normal doctrine-writing process in which one officer writes a manual and has it vetted through the senior leadership over a two- to three-year period, these processes were unusual, innovative, and swift. The vetting was aimed not at the leadership only but at the entire universe of military professionals of all ranks, as well as at academics and other professionals interested in the topic. Thus drafts were widely disseminated electronically in draft form as soon as they were close to being ready. This process—which reflects a concerted effort to provide *something* that could fill the gap in doctrine, education, and

---

58. *FM 3-07 Stability Operations* (Ann Arbor: University of Michigan Press, 2008).

training; be disseminated quickly; and then be revised based on the best available knowledge and current operational experience—demonstrates a concerted effort to institutionalize lessons learned from experience.[59]

Professional Military Education: An Engine of Change?

Perhaps the most important step in organizational learning is to ensure that lessons are passed to the next generation. Theoretically, in the military, this would be accomplished as doctrine, derived from experience as well as theory, is incorporated into education as well as training. Although officers are exposed to doctrine superficially during their precommissioning undergraduate education and in the more technically oriented first-level "basic schools" they attend early in their careers, the key injection point for operational doctrine is the midlevel service school. For the Marine Corps, this is the Command and Staff College (CSC) in Quantico, Virginia, and for the Army, it is the Command and General Staff College (CGSC) in Fort Leavenworth, Kansas. The primary purpose of this level of professional military education (PME) is to prepare officers, who have already served approximately 10 years, to serve on operational staffs and beyond.

The staff colleges can be seen as the critical educational link to doctrine in the institutional learning process for four reasons. First, it is at this point in officers' careers that they are expected for the first time to seriously read and understand doctrine.[60] It was clear through scores of interviews that this level of officer was the most well-versed in the current doctrine. Second, upon graduation, these field-grade officers are in the best position to transfer this knowledge down the chain of command. As senior majors and lieutenant colonels, they will have leadership roles as senior staff or commanders of battalions and will therefore be in charge of the next generation of leaders. Third, they represent the next generation of brigade, division, and corps commanders and senior military leaders. Because some may never attend the senior-level PME course

---

59. Lieutenant Colonel Jan Horvath, USA, personal correspondence and numerous interviews by author, October 2004–June 2005, via telephone and e-mail. More senior participants in the doctrine-writing and initial vetting process included academics such as Dr. Tom Marks; State Department representatives such as Ambassador David Passage, who had experience in El Salvador; retired Army Colonel John Wagelstein, also a veteran of El Salvador and instructor at the Naval War College; Colonel Anderson, Chief of the Special Warfare Center; and many others.

60. General Kevin Byrnes, USA, Commander, Training and Education Command, personal interview by author, February 7, 2005, Fort Monroe, VA.

(the war college). this midlevel school becomes the last place where they will be exposed in a formal way to military doctrine and concepts.

Finally, the staff colleges and especially the two advanced "graduate" schools at this level (SAMS and SAW, discussed below) are some of the most important levers each institution has to transfer new ideas *up* the chain of command. Graduates of CSC and CGSC who do not take command of a battalion will likely serve on the staff of a brigade, division, or corps commander, where they have the opportunity to share new ideas with senior leaders. Even more influential are the special graduates of the Army's School of Advanced Military Studies (SAMS) and the Marine Corps School of Advanced Warfare (SAW) who are placed on the high-level planning staffs.

SAMS and SAW are deliberately designed as a shortcut to generational learning. Following the regular 10-month staff college, the top few of the graduates (approximately 78 out of 1,000 in the Army and 24 out of 180 in the Marine Corps) are offered the opportunity to stay for an additional year to learn the operational art of military planning. Graduates of these extra-year courses are considered the elite of their generation of officers and are often referred to as the "jedi knights" of military planning.[61] The officers proceed directly upon graduation to senior-level planning staffs, where they are to practice the latest in the art of war planning. The idea is that because they are considered elite and because they are in positions to draft high-level operational plans, new ideas generated in the schools will transfer via these "jedi knights" to the senior leadership at the operational level. Thus, the organization need not wait 10 to 15 years until this generation becomes senior leaders.

This process demonstrates the way in which doctrine, in conjunction with the education and training system, can be a *driver* of change. In this top-down system, leaders can inject new ideas by changing doctrine and curricula. When working properly, these new ideas disseminate relatively quickly and change the operating paradigms of the institution. This was exactly what happened in the post-Vietnam era. With the deliberate advent of SAMS in 1982 (the Marine Corps SAW did not open until 1990), the post-Vietnam generation was able to disseminate its new doctrine

---

61. Colonel Kevin Benson, USA, Director, School of Advanced Military Planning, personal interview by author and site visit to school, March 9, 2004, Fort Leavenworth, KS; Major Isaiah Wilson, USA, "Educating the Post-Modern U.S. Army Strategic Planner: Improving the Organizational Construct" (United States Command and General Staff College, School of Advanced Military Studies, 2003).

(AirLand Battle Doctrine) relatively quickly.[62] The doctrine was also practiced in the new training center system described in chapter 4 and became reflected in the new warfighting culture.

This top-down system works well during a period of peace when the theories in doctrine are reinforced in the classroom and in exercises. When real-world operational experience conflicts with the paradigms being taught in the schoolhouses, however, officers are frustrated, and a bottom-up cycle of learning competes with the existing paradigms. This was the situation in the 1990s and in the first few years of the Iraq and Afghanistan campaigns, as waves of military officers and veterans of operations from Somalia and Haiti to Iraq and Afghanistan began to complain that the PME curricula did not reflect their experiences and that the education system was becoming irrelevant for them.

By 2004, when the first set of veterans from Iraq and Afghanistan arrived at the Army's Command and General Staff College in Fort Leavenworth, the instructors decided the students were right. The professors lacked the knowledge and curricula to provide the context needed for counterinsurgency and stability operations. To compensate, instead of lectures, they held seminars in which students shared their experience and worked out answers to common questions. In short, they began to let the students teach themselves.

In the institutional learning cycle, collecting and sharing such operational experience is a valuable step. This is not, however, the role of military education in the institutional learning process. In the military, PME provides a time for an intellectual pause where students can reflect on their experience and contemplate enduring military principles and theories. PME should provide context for those returning from a stressful and bewildering operating environment, by providing the theoretical foundations on which leaders can evaluate their experience, make sense of their challenges, and learn to make better decisions. Unfortunately, for stability operations and counterinsurgency, PME was proving increasingly unable to do this for this generation of military officers. Doctrine had been updated, but it had not been adequately incorporated into PME.

If PME fails to adapt eventually to new realities, this would reflect a "weak link" in the institutional learning cycle. A "learning institution" would respond first by recognizing this shortfall and then by updating the appropriate curricula. For true institutional learning to occur, the educational system must reflect the new lessons from the field. The ques-

---

62. Killebrew, interview; Newbold, interviews.

tion is, how soon should we expect these changes to occur in a learning institution?

Although education should adapt in response to new information, we should expect the rate of change in educational curricula to lag that of doctrine and training. This is because professional military education is supposed to reflect more enduring theoretical foundations. It cannot be instantaneously reactive to new information being sent from the field and still fulfill this role. Educators must weigh the degree to which the "fog of war" might be obscuring the more fundamental truths of modern conflict. In sum, education must be neither reactive nor reactionary but must be open to new ideas and have a process or culture that allows the identification and analysis of new information and ideas. That the instructors identified the gap in curricula and experience and doctrine indicates the beginning steps in the institutional learning cycle. Today, there is a vibrant debate among military educators about how far to "swing the pendulum" toward incorporating more COIN and stability operations into the curricula. Such debate reflects one of the key steps in an institutional learning cycle and is evidence that the system is open to change. More time is needed to determine how best to balance the need to adapt contemporaneously through PME while still retaining a focus on the enduring elements of military conflict.

Summary

This chapter has demonstrated the changes in the process of doctrine development in response to the MOOTW missions of the 1990s and the COIN and stability operations of the first decade of the 21st century. Understanding who wrote what manuals and the process by which these manuals were written sheds light on the degree to which this doctrine reflects an experiential learning cycle and organizational learning. Traditional doctrine development, with its many required senior-leader vetting processes, was designed as a top-down process. Because the actual writers of the manuals are usually midlevel officers, there is often a generational split between the writers and the sponsors of the efforts. Younger writers have often sought to align the doctrine with their own recent experience, while the older leaders may wish to reorient toward something new—or old. In the case of the marines in the Banana Wars, leadership sought to look forward to amphibious warfare versus small wars, while in the 1990s, leadership resisted peace operations doctrine in favor of the status quo, the AirLand Battle Doctrine.

From the perspective of learning theory in the military, it is important to understand the varying constraints under which the organization operates in peacetime versus war. As General Byrnes, former commander of the Army's Training and Education Command, claimed, in peacetime the institutional side of the Army "leads change," whereas in wartime, the operational side does.[63] For a military during peacetime, writing doctrine and running experiments is the only way to test new ideas and to pass on knowledge. For previous generations, this was the modus operandi for most of their careers. General Kevin Byrnes claimed that compared to officers today, his Cold War generation had more time to read the manuals.[64] It is during peacetime, not war, that officers are more likely to read the formal published doctrine. Still, without a real battlefield to test their ideas, the "experiential learning" for these peacetime warriors is based only on experimentation and training.

Once troops are sent to real combat, the battlefield becomes the new "laboratory," and the process of doctrine development needs to shift. Historically, powerful cultural resistance has often stymied this shift. Yet after years of struggling in Iraq and Afghanistan following a decade of controversy over peace operations, this resistance began to break down. This time, because the leaders reflected a generation of peacekeepers with shared experiences, doctrine writers were able to stretch the traditional system to incorporate new lessons from current operations and disseminate them more rapidly than ever before. More time is needed to determine the degree to which these contemporaneous lessons will be incorporated into core curricula in PME and institutionalized for future generations.

---

63. Byrnes, interview.
64. Ibid.

CHAPTER 6

Learning to Surge in Iraq

On January 9, 2007, President George W. Bush addressed the nation. For the first time in nearly four years of war, the president admitted, "It is clear that we need to change our strategy in Iraq." Indeed, by all accounts, the situation was deteriorating. Over 3,000 Americans had been killed and over 20,000 wounded. Accounts of Iraqi civilians killed varied from 50,000 to 75,000; and the number of internally displaced Iraqis was thought to be approaching 750,000.[1] The number of Iraqi civilians being brutally tortured and killed each night had reached staggering numbers, as sectarian violence spun out of control. The White House's "summary briefing slides" on the topic observed that "the situation in Baghdad has not improved despite tactical adjustments."

January 2007 marked the significant turning point in the Iraq War. Amid great controversy and calls to bring the troops home, President Bush announced a change in strategy, a "surge" in troop numbers, and a new commander to lead the way. Within a year, General David Petraeus had managed to stabilize Iraq and set the conditions for eventual withdrawal of combat forces. This dramatic turnaround was due not only to the deft leadership of General Petraeus but also to the existence of the new learning culture in the American armed forces—a learning culture that, as described in previous chapters, had evolved since the Vietnam era. Still, it was four years after the initial invasion before results on the ground began to be realized during the so-called Surge. For the first three years of the Iraq campaign, the coalition forces involved made little apparent progress; and indeed, by the summer of 2006, they seemed to be caught in another Vietnam-like quagmire. This chapter explains

---

1. Michael O'Hanlon and Adriana Lins de Albuquerque (later Jason Campbell), "The Iraq Index," www.brookings.edu/iraqindex (begun in Fall 2003 and updated weekly).

why it took so long for the learning curve to take effect and how the military's learning process eventually facilitated the success of the Surge.

Adapting without Winning?

For students of organizational learning, the failure to make progress in the first three years of the Iraq War was perplexing. Clearly, as this book demonstrates, at the tactical level, the U.S. military was transferring and disseminating experiential knowledge from the field and adapting contemporaneously faster than ever before. In nearly every way, this process reflects the bottom-up organizational learning model prescribed by John Nagl in his book *Learning to Eat Soup with a Knife*. From doctrine revision (including the production of an all-new and widely disseminated counterinsurgency manual) to massive adaptation in training, education, in-theater schools, and online resources, this generation exhibited a remarkable ability to learn in the field and adapt contemporaneously.[2] This learning curve was a product of the dynamic "get-the-ground-truth" learning processes created and honed in previous decades and described in detail in the previous chapters. The obvious question is, why, after years of learning and adapting, was the U.S. still struggling to win the peace in Iraq?[3]

The first part of this chapter examines this puzzle, highlighting the clear evidence of military learning and adaptation, as well as the limitations of the current learning system as applied to the complex civil-military conflict in Iraq. It reveals that while troops at the tactical and operational level certainly benefited from the new and improved learning processes described here, *military* learning proved necessary but not sufficient for achieving success, given the complex civil-military factors at play.

Three factors in the Iraq case complicated strategic success in the early years. First, a series of horribly uninformed choices at the strategic and political levels, combined with poor assumptions made by (and forced on) war planners and civilian administrators, made an incredibly challenging set of initial conditions even more difficult. Second, ironically, the parallel process of top-down military theorizing and experimenting on the topic of interagency operations throughout the 1990s led to unrealistic expectations about the capabilities and capacities of

---

2. David Ucko, "Innovation or Inertia: The U.S. Military and the Learning of Counterinsurgency," *Orbis* (Spring 2008).

3. "Iraq: What Next," special issue, *New Republic* (2006); O'Hanlon and Campbell, "The Iraq Index."

other nonmilitary government agencies in the planning phases and beyond. Third, the sheer scale and strategic nature of the complex stabilization and reconstruction operations in Iraq presented unique challenges to the tactically oriented military learning system.

To meet the challenges of 21st-century conflict, this learning system will need to be adapted and enlarged to incorporate the other, nonmilitary elements of the U.S. government involved in stabilizing complex societies. This process will be complicated by the lack of a learning culture in civilian agencies, which is the key prerequisite to a process of getting to the "ground truth." Finally, it is important to note that conflicts such as that faced in Iraq have historically taken over a decade to resolve. Thus both the final outcome of the Iraq campaign and the effect the campaign will have on the orientation of the U.S. military will take years, perhaps a generation or two, to become evident.

Setting the Stage for What Went Wrong: Failures in Strategic and Political Leadership

The case of Iraq demonstrates clearly how learning at the tactical level does not necessarily aggregate to strategic success. Decisions made by civilians as well as military leaders at the strategic and political levels have a profound effect on the conditions under which the military operates. In Iraq, poor civilian and military decision making at the strategic and political levels—from prewar planning to occupation and beyond—limited tactical and operational options, confused and frustrated commanders, and continually exacerbated an already complex operating environment. This demonstrates the limits of bottom-up military adaptation in overcoming poor strategic decisions made by uninformed but empowered leaders. Strategic coherence is necessary in order to guarantee that commanders on the ground have the right resources, the right message, and the political legitimacy they need to carry out their mission.

Misunderstanding Initial Conditions and the Decision to Invade

Perhaps the first, most egregious error in strategic judgment made in the case of Iraq was the decision to invade in the first place. Invading and occupying a country as complex and troubled as Iraq would have been difficult enough even with enlightened leadership, improved bottom-up military learning processes, and increased capability and capacity on the civilian side of government. Initial conditions in Iraq should have been a warning to savvy planners and political leaders that the lofty goals of regime change could not be achieved through swift military defeat of a third-rate army and disregard for post-invasion planning.

Decades of tyranny by Saddam Hussein had resulted in a deeply damaged country—physically, economically, and psychologically. Iraq's infrastructure, from its electrical grid to its health care system, was in grave disrepair. The country had the highest infant mortality rate and the shortest life expectancy in the region. The state-run economy was overly reliant on oil production and uncompetitive in a global market. Most critically, ethnic tensions had been held at bay through brutal repression of the Shia majority by the Sunni minority, while no-fly zones, enforced by American air power, had provided sanctuary for the Kurdish population for over 10 years. In a post-Saddam Iraq, Shia would have high expectations for justice vis-à-vis the Sunni, and Kurds would expect sovereignty as well as liberation following the American invasion. Meanwhile, Iraq's complex relations with its regional neighbors, including Syria, Saudi Arabia, Iran, and Kuwait, remained strained by transnational ethnic ties, economic competition over oil, and still vivid memories of bloody wars waged in previous decades. To most informed observers, these conditions were a tinderbox for bloody reprisals, civil war, and regional intervention.

Any plan to invade, occupy, and "fix" Iraq should have taken into consideration these preexisting conditions. One might speculate that a clear understanding of the challenges these factors presented to an invading force might even have been a deterrent to military invasion in the first place. In fact, then secretary of defense Dick Cheney made a similar argument against ousting Saddam Hussein after the First Gulf War in 1991.

> How long would we have had to stay in Baghdad to keep that government in place? What would happen to the government once U.S. forces withdrew? How many casualties should the United States accept in that effort to try to create clarity and stability in a situation that is inherently unstable?[4]

But despite myriad warnings and studies issued by bureaucrats and analysts inside government as well as academics and pundits on the outside, decision makers in Washington proceeded to direct plans based on best-case assumptions: coalition forces would be treated as liberators;

---

4. Secretary of Defense Richard Cheney, "The Gulf War: A First Assessment" (paper presented at the Soref Symposium, Washington, DC, Washington Institute for Near East Policy, 1991).

Iraq's oil economy would fund postwar reconstruction; a free democratic Iraq would thrive in its new market economy and be a beacon of stability in a volatile political region. In short, as Ken Adelman famously claimed in his 2002 *Washington Post* op-ed, regime change in Iraq would be a "cakewalk."[5] This overconfidence and failure in planning has been widely documented in numerous books and articles and can be seen as the "first sin" of the Iraq War.[6]

Iraq Planning and the American Military Culture

Planning failures in Iraq have been chronicled in too much detail elsewhere to be repeated here.[7] However, with respect to learning and the U.S. military, there are a few points to be highlighted and assertions that need to be set straight. First, the failures to plan accordingly for the postwar period (so-called Phase IV) are seen by many as evidence that the U.S. military was still stuck in its post-Vietnam mind-set—meaning that it not only had forgotten any potentially relevant lessons from its Vietnam experience but had created a culture in which the very notion of conducting such nontraditional military missions could not have been contemplated or understood.[8] This "cartoon" image of the American military culture ignores the evolution that had taken place during the 1990s in doctrine and training for peace operations and the substantive lessons that had been learned by a generation of officers who had cut their teeth in the streets of Somalia, Haiti, and the Balkans.

In reality, as a result of the mismatch between training and education, on the one hand, and operational experience, on the other, throughout the 1990s (as discussed in previous chapters), various competing perspectives existed among U.S. military officers and across generations with respect to noncombat roles and the nature of modern conflict. These perspectives informed the debate in the Iraq planning phases

---

5. Ken Adelman, "Cakewalk in Iraq," *Washington Post*, February 13, 2002.

6. For detailed accounts of the many warnings and failures in planning, see the following: Michael Gordon and Bernard Trainer, *Cobra II: The Inside Story of the Invasion and Occupation of Iraq* (New York: Pantheon, 2006); George Packer, *Assassin's Gate: America in Iraq* (New York: Farrar, Straus and Giroux, 2005); Thomas Ricks, *Fiasco: The American Military Adventure in Iraq* (New York: Penguin Press, 2006); James Fallows, "Blind into Baghdad," *Atlantic* (January–February 2004); Joseph J. Collins, "Choosing War: The Decision to Invade Iraq and Its Aftermath" (National Defense University Occasional Paper, April 2008).

7. Collins, "Choosing War"; Fallows, "Blind into Baghdad."

8. General Jack Keane, quoted by John Nagl in "Counterinsurgency in Vietnam: American Organizational Culture and Learning," in *Counterinsurgency in Modern Warfare*, ed. Daniel Marston and Carter Malkasian (London: Osprey, 2008).

about if and how to conduct postwar "cleanup operations" and whether coalition forces would likely face an insurgency. The colonels who comprised the Central Command (CENTCOM) planning staff were of the generation whose operational experience began at the end of the Cold War and consisted primarily of "other-than-war" contingencies. According to Colonel John Agoglia, one of the key planners for CENTCOM, the planning team did contemplate the need to conduct stability operations and potentially a counterinsurgency following the "major combat" phase. Their assessments indicated that the numbers of troops that would have been required to secure cities along the way and in Baghdad after the invasion would have been immense.

When the planners queried Washington about these post-invasion issues and the need for large numbers of follow-on troops to stabilize the country, the daunting realities of providing law and order across a chaotic postwar society were effectively wished away at the highest levels of leadership with the assumption that the American forces would be greeted as "liberators," not "occupiers." Thus, according to Agoglia, there could be no plan to secure Baghdad because of resource constraints placed on CENTCOM planners from Washington and the failure of the more senior uniformed military leadership to push back sufficiently.[9] When Deputy Secretary of Defense Paul Wolfowitz publicly ridiculed the Chief of Staff of the Army, General Eric Shinseki, for suggesting that it would take more troops to occupy and stabilize a post-invasion Iraq than it would to topple the regime, the military leadership got the message.

ECLIPSE II, the postwar plan eventually developed by Colonel Kevin Bensen before the invasion, called for over 300,000 troops to stabilize the country following the invasion.[10] This ratio of 11 troops per 1,000 residents reflected accepted "best practices" for counterinsurgency and demonstrated that planners did have an understanding of the requirements of the messy postwar environment they were likely to encounter. Still, the estimate was wildly out of touch with what would be politically acceptable, given the fight over troop numbers the staff had already been through with Secretary Rumsfeld in previous months with respect to the major combat phase.

Three additional assumptions in ECLIPSE II would further under-

---

9. Series of personal interviews with Colonel John Agoglia, U.S. Army, 2005–8.
10. Donald P. Wright, Colonel Timothy Reese, and the Contemporary Operations Study Team, *On Point II: Transition to a New Campaign—The United States Army in Operation Iraqi Freedom, May 2003–January 2005* (Combat Studies Institute Press, US Army Combined Arms Center, 2008).

mine success for troops on the ground. First was the assumption that there would be adequate Iraqi infrastructure in place to facilitate post-combat operations. Second, planners assumed, due to faulty intelligence reports, that there was little risk of a large-scale insurgency. Third, as one colonel involved in the planning claimed, "we made an assumption in the original OPLAN that there would be some level of [Iraqi] security forces, both Army and police, that could be leveraged to provide immediate local security and that it would form a core for the rebuilding of an Iraqi Army." All of these assumptions proved woefully off the mark.[11] Evidently there were no branch plans incorporated into the base plan to account for the possibility that these assumptions would prove false.

Although a more comprehensive assessment of existing data from available experts would have informed these military planners of the decrepit state of the Iraqi infrastructure, the other two assumptions were a bit more understandable. One could make the case that had the available Iraqi security forces been leveraged, the wide-scale resentment toward the coalition and the ensuing insurgency might have been prevented. However, the first and second official acts of Paul Bremer, head of the Coalition Provisional Authority (CPA)—CPA Order Number 1 and CPA Order Number 2—respectively gutted the bureaucracy and dissolved the Iraqi Army; hence the coalition's own leadership ensured that this was not to be the case.[12]

Attempts by the coalition to make up for the poor Phase IV planning were severely undermined by poor political decisions made by the CPA. Building an Iraqi army and a civil bureaucracy from the ground up would prove much more difficult than leveraging existing talent and structure might have been. On the ground, the effect was visceral. Colonel Alan King told journalist Tom Ricks, "When Bremer did that, the insurgency went crazy . . . One Iraqi who had saved my life in an ambush said to me, 'I can't be your friend anymore.'"[13]

Although there is much blame to be assigned to both the military and civilians in the planning phase, CPA Order Number 1 and CPA Order Number 2, combined with the refusal to deploy adequate numbers of troops, clearly added fuel to the fire, making an extremely challenging mission even more difficult. Had Benson's plan not been undermined by CPA decision making and had he been assured an adequate number of

---

11. Wright, Reese, and the Contemporary Operations Study Team, "On Point II."
12. Paul Bremer, "Coalition Provisional Authority Order Number 2: Dissolution of Entities" (Baghdad, May 23, 2003); Paul Bremer, "Coalition Provisional Authority Order Number 1: De-Ba'Athification of Iraqi Society" (Baghdad, May 12, 2003).
13. Colonel Alan King quoted in Ricks, *Fiasco,* 164.

troops for stabilization, the rebuilding of Iraq might have stood a better chance. As it was, proficiency and rapid adaptation and learning at the military or tactical level would not easily compensate.

Learning in a Vacuum: The Myth of "the Interagency"

Another factor that complicated success in Iraq was the commonly held theory among military personnel that deployable experts who are available and can conduct the myriad stabilization and reconstruction tasks needed to ensure political success in the aftermath of an invasion exist in the State Department, the U.S. Agency for International Development (USAID), the Treasury, the Justice and Commerce departments, and even the Department of Agriculture. While the military does its job fighting the enemy, it is assumed that these other, nonmilitary partners will arrive to win the peace. It is an elegant theory, but one that is unfortunately not based on historical experience or supported by the capability and capacity of the U.S. government. These unrealistic expectations about the capacity and capability of nonmilitary agencies and partners undermined success in Iraq from planning to execution and beyond.

The capability and capacity of the so-called interagency (the military's shorthand for all nonmilitary U.S. government agencies) is simply dwarfed by that of the U.S. military. The Army's basic deployable unit, a brigade combat team (BCT), can be as large as 4,000 troops. There are approximately 40 BCTs in the U.S. Army. By contrast, there are fewer than 2,000 USAID officers in the entire agency, and they are deployed around the entire world. The State Department is not much better off, with 6,000 foreign service officers (FSOs) on its payroll, whose primary job is to serve as diplomats at over 265 posts around the world.[14] The running joke among civilian and military U.S. government bureaucrats is that there are more lawyers in the Department of Defense (approximately 11,000) than there are USAID and foreign service officers combined.

Aside from the pure numbers gap, there are also myths about core competency. FSOs are diplomats. They are no more trained for kickstarting governments and conducting conflict resolution than are military commanders. Similarly, although USAID officers do possess more knowledge of the nuances of economic development, they do not usually perform these tasks on the ground themselves. USAID officers con-

---

14. U.S. Department of State, Bureau of International Information Programs, www.state.gov.

tract with local and international partner organizations and businesses who actually perform the work. Contracting with locals is designed to help the local economy get on its feet.

USAID's economic development methods are well honed, and USAID officers are experts in contracting for a desired end state. They understand how to do this in ways that promote long-term economic development objectives while mitigating unintended consequences. One enlightened military officer noted, "There's a reason that USAID has a 5-foot-tall book on regulations for contracting."[15] Unfortunately, while this model works well in relatively peaceful developing nations, it runs into trouble in the violent postwar and counterinsurgency scenarios where unarmed aid workers are vulnerable and contracting partners often refuse to operate.

Given this lack of capability and capacity in civilian agencies, the question is, why did military officers assume that when they marched into Iraq, these responsibilities would be taken care of by "the interagency?" The answer is that it is what they were taught. Ironically, just as a generation learned AirLand Battle Doctrine through the top-down scientific method of theorizing, experimenting, and disseminating throughout the 1980s, so this generation taught itself a new theory of civil-military operations throughout the 1990s. Unfortunately, this theory was just a theory. It was not supported by the realities in the other government agencies, nor did it have strong recent or historical experience as its guide. Misleading doctrine on this topic—developed through a series of after-action exercises and theorizing since the end of the Cold War—contributed to this misperception.

An often-cited historical model for civil-military coordination is the U.S. military's experience with the Civilian Operations Revolutionary Development Support (CORDS) in Vietnam. CORDS was a combined civilian and military effort in which multiple civilian agencies in the U.S. government, including USAID, the CIA, the State Department, and the U.S. Information Agency, coordinated their missions with the U.S. military to achieve "unity of effort." A special vignette is dedicated to this example in the DoD's joint publication on interagency (JP 3-08), published in the mid-1990s. The manual claims, "In a broad context CORDS provides an excellent example of an effective campaign plan within an interagency context."

Unfortunately, the severe lack of capability and capacity in civilian

---

15. Corine Hegland, "National Security—Why Civilians Instead of Soldiers?" *National Journal,* April 28, 2007.

agencies would make application of any CORDS-type program in the post–Vietnam era unrealistic. At its peak during the Vietnam War, US-AID had approximately 15,000 officers in theater—still only about three-fourths the size of one Army infantry division. By the 1990s, just as the military was rediscovering that nonmilitary partners might be "force multipliers" (a label these partners would not chose), U.S. government agencies such as USAID were having their budgets slashed by Congress. The fact that this theory of civil-military coordination was able to take hold in the 1990s after USAID had been reduced to a small cadre of contracting professionals reflects the pathology of the military's top-down scientific method for preparing for conflict, even as it operates alongside the tactical-level learning process described in this book.

The modern roots of this theory began with the ill-fated episode in Somalia, after which the military began to try to understand the nonmilitary actors they had encountered there. Through a series of well-intentioned workshops and exercises, both the Marine Corps and the Army made a concerted effort to improve civil-military operations by capturing lessons from real operations and reaching out to the NGO and interagency communities to find ways to improve communications and, in some cases, coordinated operations. In 1995, General Anthony Zinni, who felt that the biggest problems in Somalia had stemmed from civil-military coordination, focused an existing Marine Corps exercise, called "Emerald Express," on the civil-military problem at the operational level.[16] Under his leadership as the commander of the First Marine Expeditionary Force and, later, as the commander of CENTCOM, Emerald Express became a series of weeklong operational-level workshops designed to share lessons among the civil-military community.[17] NGOs, foreign militaries, and U.S. government agencies, as well as representatives from the Army's Peacekeeping Institute and the First Marine Expeditionary Force, participated. U.N. ambassador Madeline Albright even gave the keynote address at the first conference in April 1995.[18]

More than just an information-sharing or lessons-learned after-action forum, Emerald Express allowed practitioners to identify and collectively

---

16. General (Ret.) Anthony Zinni, USMC, personal interview by author, January 31, 2005, Arlington, VA, http://www.jfcom.mil/about/experiments/mc02.htm, http://www.jfcom.mil/about/glossary.htm#JE.

17. Sandra Newett et al., *Emerald Express '95: Analysis Report* (Alexandria, VA: Center for Naval Analysis, 1996).

18. General (Ret.) Tom Clancy, Anthony Zinni, USMC, and Tony Koltz, *Battle Ready* (New York: Putnam, 2004); Rod Deutschmann, "Top Military, Civilian Officials Discuss Future of Humanitarian Operations," *Navy Wire Service*, Navy Public Affairs Library, April 27, 1995; Zinni, interview.

solve operational problems that had been encountered in real-world scenarios. Problems were discussed, and the solutions derived during the conference were published in pamphlets and distributed to the major commands and services. Thus, throughout the 1990s, this community began to hone in on an improved model for interagency, civil-military coordination. They were working to change the system by collectively determining how they wanted things to be.

The problem was that the ideas developed in this process were prematurely fed into doctrine, training, and education. Not only was General Zinni interviewed by the doctrine writers and quoted directly in JP 3-08 *Interagency Coordination during Joint Operations*, but many of the doctrinal concepts outlined in the manual had been formulated during Emerald Express and published in pamphlets previously. As this doctrine began to find its way into education, experimentation, and training, a new generation of military leaders began to learn new and improved methods for coordinating civil-military operations.

The focus of this batch of doctrine was on understanding the cultures and structures of other actors and on the many processes by which a commander might coordinate with them. There is a consistent emphasis on the theme that the military acts *in support of* civilian governmental actors. For example, in its discussion on peace operations, the Army's 2001 FM 3-07 *Stability Operations and Support Operations (SASO)* states, "While these activities are primarily the responsibility of civilian agencies, the military can support these efforts within its capabilities."

To the operationally oriented military mind, saying that an agency has "primary responsibility" is easily interpreted to mean that these agencies will *conduct* these tasks with or without support from the military. There is no significant discussion about what "support to" such agencies by the military might actually entail. From the State Department's perspective, "responsibility" for such operations might entail setting the objectives and providing guidance, with the expectation that the military would have a significant role in actually providing security and, in many cases, conducting the tasks. The State Department has no expeditionary capacity of its own.

Such assumptions are further reflected in the annex of volume 2 of JP 3-08, published in 1995, where a chart identifies the capabilities of other agencies and NGOs (e.g., "food and water," "sanitation," "clothing and medicine," "refugee services," etc). Nowhere, however, is there a discussion of the actual operational *capacities* of these agencies. Thus, troops' assumptions about what other agencies and NGOs would bring to an operation are somewhat understandable. This generation of military officers was taught that nonmilitary agencies have the capability and responsibility—indeed, the lead—for these tasks. The fact that these other agencies did not have the capability or capacity—or, in the cases of many NGOs, the *will*—to conduct these tasks in partnership with the military (especially while the bullets were still flying) did not seem to occur to them.

This oversight by the doctrine writers is even more perplexing in light of U.S. experience in Haiti in 1994. In an after-action report conducted by the National Defense University, problems with civilian capacity were clear.

> The lack of civilian surge capacity puts civilian agencies at a disadvantage in contingency planning and, in Haiti, delayed their ability to bring resources to bear in the initial days following the Haiti intervention. The military, in contrast, expected the civilian agencies to come with resources in hand. Without a dedicated planning cadre and some surge capability developed within civilian agencies, or relying on the military, efficient interagency planning and coordination will remain illusive.[19]

Assumptions expressed in these manuals were based on theory rather than experience and were misleading with respect to the division of labor and what other agencies could realistically provide. The military had discovered "partners," and they expected them to perform. In short, the ideas reflected how this community—both civilian and military—*wanted* things to work, not how they actually could work given the current capability and capacity. Serious policy changes and vast resource allocations to grow civilian agencies would need to be made in order to realize this vision, none of which would have been made by the time coalition forces marched into Baghdad.

In Iraq, these misperceptions about civilian agencies have had grave consequences. Massive gaps were evident in the overall effort to get Iraq running economically and politically following the invasion. Not only was the scale of operations in Iraq far greater than anything U.S. nonmilitary agencies had addressed in the 1990s, but the level of violence challenged civilians' institutional ability to be there at all. In short, these agencies had neither the capacity nor the capability to provide the partnership that the U.S. military expected. As a result, many required tasks, such as developing a comprehensive "whole-of-government" reconstruc-

---

19. Margaret Daly Hayes and Gary F. Weatley, National Institute of Peace Studies, *Interagency and Political-Military Dimensions of Peace Operations: Haiti, a Case Study* (Washington, DC: National Defense University Press, 1996), chap. 5.

tion and economic development strategy or a comprehensive legal infrastructure (including prisons, courts, and the rule of law), were not adequately accomplished and integrated into the military's strategic plan.

Despite the evidence that nonmilitary agencies lacked the capability or capacity to operate in Iraq, the military continued to cling to the false theory from 2003 onward. Many military and civilian leaders claimed that the number one problem plaguing the U.S. military's efforts in Iraq was the lack of civilian U.S. government partners. In the spring of 2007, young officers interviewed in the media expressed frustration that the only part of the U.S. government that was "surging" for General Petraeus's new strategy was the military. Indeed, when President Bush appointed Lieutenant General Doug Lute as "war czar" in May 2007, this experienced leader quickly asserted his goal of getting the nonmilitary agencies to "step up" to take over the so-called political and economic lines of operations.[20]

Lieutenant General Lute's vision of a "whole-of-government" approach echoed statements by the secretary of state Condoleezza Rice when she promoted the theory in her congressional testimony in January 2007.

> Success in Iraq, however, relies on more than military efforts alone; it also requires robust political and economic progress. Our military operations must be fully integrated with our civilian and diplomatic efforts, across the entire U.S. government, to advance the strategy that I laid out before you last year: "clear, hold, and build." All of us in the State Department fully understand our role in this mission, and we are prepared to play it. We are ready to strengthen, indeed to "surge," our civilian efforts.[21]

In this statement, Secretary Rice writes a check her agency clearly cannot cash as she promises "to deploy hundreds of additional civilians across Iraq to help Iraqis build their nation." These civilians, she claims, will man the civilian slots in the 18 provincial reconstruction teams (PRTs), which are designed to work with Iraqis at the local level to promote economic and political development. Despite the clear lack of capacity, such statements by both the 2007 "war czar" and the secretary of state perpetuated the myth that civilian agencies might be able to stand up to the responsibilities assigned to them in this civil-military theory of operations.

---

20. Peter Baker and Robin Wright, "To 'War Czar,' Solution to Iraq Conflict Won't Be Purely Military," *Washington Post,* May 17, 2007.

21. House Committee on Foreign Affairs, *Congressional Testimony of Secretary of State Condoleezza Rice: The New Way Forward in Iraq,* January 11, 2007.

Within weeks of the secretary's announcement, the State Department was forced to turn to the Pentagon for volunteers to fill these "civilian" PRT posts.[22]

The lessons learned about interagency operations throughout the 1990s reflected the military's scientific top-down learning method and a misunderstanding of the applicability of historical models given current agency structures. The ideas that were developed, written into doctrine, and disseminated were based on the judgment and expertise of many prominent civilian and military participants; but they still only reflected a *theory* about how they *wanted* things to work in the field, not necessarily actual experience in nonpermissive environments. Even more problematic, these theories were not measured against the actual capacity and capability of nonmilitary agencies.

Unfortunately, the military's learning system was unable to compensate for this interagency theorizing. The military's system, as described in this book, is a military system and was not designed to collect lessons from other agencies. The military's lack of knowledge of these agencies was evident—and attempts to learn and update military doctrine were often stymied by the fact that there were too few civilians available to inform their efforts. Contractors, not real agency actors, were (and still are) often hired to "role-play" as stand-ins for interagency actors in military experimentation, exercises, and training. This lack of contact with the actual agency personnel often served to confirm, rather than challenge, the assumptions being made in doctrine development, as the actors often perform as prescribed by the published doctrine. The fact that the State Department or USAID lacked personnel to participate in these exercises should have been noted as an indication that they might not have the capacity to participate in real life.

The latest iteration of the Army's field manual for stability operations, FM 3-07 (published in the fall of 2008), reflects progress in this area. Doctrine writer Lieutenant Colonel Steve Leonard, under the leadership of Lieutenant General William Caldwell, General Petraeus's successor at the Combined Arms Center, led an innovative process for the development of this manual. Recognizing the pitfalls of writing untested theory into doctrine, he hosted a series of roundtables from 2007 to 2008 with midlevel civilian agency and NGO practitioners to vet the language in the book. His specific questions sought to determine current

---

22. Robert Perito, "Provincial Reconstruction Teams in Iraq," U.S. Institute of Peace briefing, http://www.usip.org/pubs/usipeace_briefings/2007/0220_prt_iraq.html (February 2007).

fact from "aspirational" theory and to identify areas where one agency might interpret language differently from another. The result was a manual that seems to more accurately reflect nonmilitary realities.[23]

Learning to "Surge"

Despite challenges in leadership and planning and false expectations vis-à-vis interagency capabilities, the U.S. military's bottom-up learning curve was swift in Iraq. Leveraging the institutional processes described in this book, military leaders were able to rapidly identify problems and lessons and disseminate new tactics faster than ever before. The most obvious indication of military adaptation and learning in Iraq—and the most clear example of the limitations of that tactically oriented system in promoting strategic success—was the change in strategy adopted in the spring of 2007, the so-called Surge.

The Surge formalized the approach a number of commanders had been implementing independently across different areas of the country and reflected the intellectual and doctrinal shifts that had occurred in the first years of the war. As discussed below, commanders who had attempted to implement a more COIN-oriented strategy based on the success of some of their peers and an increased understanding of COIN principles had been increasingly stymied by a lack of resources to get the job done. Thus, in addition to an increase of nearly 30,000 troops, what made the Surge different was its longer-term approach and a change in tactics that emphasized interacting with the population and protecting civilians as a key to success—and a U.S. military that was ready to shift course.

Still, not everyone bought into the need for more troops and a longer-term COIN-oriented approach.[24] General William Casey, the commander of the Multi-National Force–Iraq (MNF-I) from 2005 to 2007, had actively resisted implementing the president's "clear, hold, and build" strategy, which mirrored classic COIN approaches, and he had never requested the number of troops needed to carry it out. Casey, who was the first to develop an actual COIN-oriented campaign plan, had still attempted to keep troop levels at a lower number while decreasing the overt coalition footprint.

Both General Casey and CENTCOM commander General Abizaid pub-

---

23. Participant observation by the author in various doctrine workshops, November 2007, April 2008, October 2008, Washington, DC. See also Ann Scott Tyson, "Standard Warfare May Be Eclipsed by Nation-Building," *Washington Post*, October 5, 2008.
24. Gian Gentile, "Eating Soup with a Spoon: Missing from the New Coin Manual's Pages Is the Imperative to Fight," *Armed Forces Journal* (September 2007).

licly criticized Petraeus's proposed force increases.[25] General Abizaid stated bluntly to Congress, "I do not believe that more American troops right now is the solution to the problem. I believe that the troop levels need to stay where they are."[26] Likewise, the chairman of the Joint Chiefs of Staff, General Pace, who had led an internal assessment on Iraq strategy that also favored fewer troops, was frustrated that the White House seemed to dismiss his team's assessment that more troops were not the answer.[27]

Although Casey claimed to be conducting COIN, in contrast to the Surge strategy, Casey had focused on training Iraqi security forces and rebuilding the Iraqi Army, as his plan's primary objective was to hand over the mission to Iraqis and redeploy as soon as possible. While most would agree that this was a laudable goal, the effort had been hampered by increasing levels of violence among Iraqis, a growing Al Qaeda presence, and heightened sectarian conflict. Retreating to large forward operating bases (FOBs) and leaving population security to inexperienced Iraqi security forces, many of whom were being pulled into the sectarian conflict themselves, was simply not working to reduce overall levels of violence. Violence needed to be brought under control so that Iraqis could feel safe before retreating to FOBs would make sense.

This mismatch between Casey's approach for swift transition and redeployment and the White House's published 2005 *Strategy for Victory in Iraq*, which outlined a "clear, hold, and build" approach, created strategic ambiguity on the ground. The net effect was a patchwork of different approaches being tested in the field at the brigade and division levels. Some commanders, such as H. R. McMaster, a brigade commander in Tal Afar, and Peter Chiarelli, a division commander in Baghdad, had attempted population-focused COIN strategies in their areas of responsibility and had had widely publicized success.[28] But a lack of troops to clear, hold, and build, combined with Casey's push to have U.S. troops reduce their presence and retreat to large FOBs, made the widespread application of similar approaches in other areas increasingly untenable.

---

25. Ray Suarez, "Gen. Casey Faces Criticism in Senate Confirmation Hearing," *News Hour*, PBS, February 1, 2007, http://www.pbs.org/newshour/bb/military/jan-june07/casey_02-01.html.

26. Peter Baker, "President Confronts Dissent on Troop Levels: Bush Indicates Military Won't Dictate Numbers; Top General to Retire," *Washington Post*, December 21, 2006, A01.

27. Bob Woodward, "Outmaneuvered and Outranked, Military Chiefs Became Outsiders," *Washington Post*, September 8, 2008.

28. George Packer, "The Lesson of Tal Afar," *New Yorker*, April 10, 2006; Peter Chiarelli and Patrick R. Michaelis, "Winning the Peace: The Requirement for Full Spectrum Operations," *Military Review* (July–August 2005).

So, although commanders were sharing success stories and learning what worked and what did not, application of the actual practice was stymied by decisions at higher levels that limited access to resources needed to get the job done.

Organizational learning theorists might recognize this phase as a necessary step in institutional learning. Paradigms must be challenged, and a period of frustration and debate is what often sparks the learning cycle to progress. In this case, although the evidence of learning from the field was clear among many (not all) of the commanders at various levels in Iraq, many at the top of the institution—including the secretary of defense—were still not convinced. COIN-oriented commanders on the ground were sharing their experience throughout the system, yet they lacked the resources and strategic guidance at the top of the organization that would enable them to implement a comprehensive COIN strategy throughout the country. Bottom-up contemporaneous adaptation might have been necessary but was not sufficient. The most senior leadership had not yet come to consensus, and many had yet to make the cognitive shift.

Advocacy and the Intellectual Foundations of the Surge

The intellectual foundations for the Surge began with debates among scholars, pundits, and practitioners as early as 2005. Studies by think tanks such as the American Enterprise Institute (AEI), the Brookings Institution, and the U.S. Institute of Peace competed for the president's attention with internal assessment conducted by the chairman of the Joint Chiefs of Staff.[29] One of the first to publicly call for a complete overhaul in strategy was Dr. Andy Krepinevich, a retired Army colonel and counterinsurgency expert at the Center for Strategic and Budgetary Analysis. In September 2005, Krepinevich published an influential article for *Foreign Affairs* in which he criticized the administration for not having a strategy in Iraq at all.

> As President George W. Bush has stated, "Our strategy can be summed up this way: as the Iraqis stand up, we will stand down." But the president is describing a withdrawal plan rather than a strategy.

---

29. James A. Baker and Lee H. Hamilton, *The Iraq Study Group Report* (Washington, DC: United States Institute of Peace, 2006), http://www.usip.org/isg/iraq_study_group_report/report/1206/index.html; Kenneth Pollack, "A Switch in Time: A New Strategy for America in Iraq" (Brookings Institution, February 2006); Fred Kagan, *Choosing Victory: A Plan for Success in Iraq, Phase I Report* (Washington, DC: American Enterprise Institute, 2006).

Krepinevich advocated a classic "oil spot" approach to stabilize Iraq, "in which operations would be oriented around securing the population and then gradually but inexorably expanded to increase control over contested areas."[30] His assessment drew from historical cases and classic COIN theory. Similar themes were reflected in the Brookings Institution report published in February 2006, which called for a strategy that would "make protecting the Iraqi people and civilian infra-structure its highest priority, training Iraqi security forces a close second, and hunting insurgents a distant third."[31]

At the White House, as violence continued to escalate through 2006, one think tank in particular stood out. As former deputy national security adviser Meghan O'Sullivan explained in a *Washington Post* online chat, a critical push for a change in strategy came from these external analysts.

> Some of the scholars at AEI played a number of very important roles in the shift in strategy. Fred Kagan came and, with other outside scholars, spoke to President Bush at Camp David in June 2006—which was an early opportunity to air views that were outside conventional thinking on Iraq and helped ensure that the debate would consider all the options as it matured. But perhaps among the most useful role played by AEI scholars was their ability and willingness to be public advocates for a move—the addition of more troops into Iraq—that was nearly universally unpopular in the country at the time . . . There were active debates inside the administration and internal supporters of such a strategic shift. But these people were not in a position to discuss such views externally. Other outsiders also played important roles, by providing different perspectives, offering candid assessments, and challenging conventional thinking.[32]

In January 2007, as the president was announcing his new "surge" strategy, AEI published the results of their Iraq Planning Group study, a 50-page report by Fred Kagan and retired Army general Jack Keane titled *Choosing Victory: A Plan for Success in Iraq*.[33] Their more optimistic assessment and their assertion that there was still a "path to victory" had clearly resonated with President Bush, who was not ready to admit defeat.[34]

---

30. Andrew Krepinevich, Jr., "How to Win in Iraq," *Foreign Affairs* (September–October 2005).

31. Pollack, "A Stitch in Time."

32. Transcript from September 15, 2008, http://www.washingtonpost.com/wp-dyn/content/discussion/2008/09/12/DI2008091202739.html?sid=ST2008090404206&s_pos=list.

33. Kagan, *Choosing Victory*.

34. Bob Woodward, *The War Within: The Secret White House History 2006–2008* (New York: Simon and Schuster, 2008).

The Kagan-Keane study differed from emerging popular opinion at the time that, in the face of escalating violence and a rising U.S. death toll, it was time to start cutting losses and draw down. The latter perspective was held by various pundits and was also reflected in the widely publicized, congressionally sponsored, bipartisan Baker-Hamilton Commission's *Iraq Study Group Report,* facilitated through the U.S. Institute of Peace. This report called for an increased emphasis on training the Iraqi Army and suggested that it could be accomplished swiftly: "We should seek to complete the training and equipping mission by the first quarter of 2008."[35] On the question of increased troop deployments, the report was unambiguous: "Sustained increases in U.S. troop levels would not solve the fundamental cause of violence in Iraq, which is the absence of national reconciliation."[36]

In contrast, the AEI study claimed that there might still be a "path to victory," but only with a significant change in military and political strategy and a longer-term commitment. Although many observers focused on the fact that the new approach called for a significant increase in troops, the so-called Surge of 2007 was much more than a quantitative change. What made the Surge significant was the *qualitative* shift in strategy—from a focus on training Iraqi security forces while also targeting the enemy directly ("enemy-centric") to an emphasis on protecting Iraqi civilians in the most populous city, Baghdad, in an effort to marginalize the enemy ("population-centric"). This was designed to reduce violence to a level in which the political reconciliation identified by Baker-Hamilton could take place. Moreover, Kagan and Keane emphasized that this was not a "quick fix" strategy: due to the high level of violence and the logistical requirements to move new forces to theater and reposition them out among the population, it would require at least 18 months before the results materialized.[37] Finally, General David Petraeus, the charismatic leader who had written his dissertation on counterinsurgency and had just completed the redrafting of the new counterinsurgency doctrine, was given the command.

Doctrinal Foundation for Implementation of the Surge

In the fall of 2005, over two and a half years into the struggle to stabilize Iraq, General Petraeus had partnered with Marine Corps general James

---

35. Baker and Hamilton, *The Iraq Study Group Report.*
36. Baker and Hamilton, *The Iraq Study Group Report,* 51.
37. Fred Kagan and Jack Keane, "The Right Type of 'Surge': Any Troop Increase Must Be Large and Lasting," *Washington Post,* December 27, 2007.

Mattis to lead a yearlong effort to rewrite the Army and Marine Corps counterinsurgency field manual (FM 3-24). This new doctrine drew heavily on their own experiences and on historical case studies and "best practices" of past counterinsurgencies. The final draft of the manual was published in the fall of 2006, but previous drafts had been distributed liberally via email throughout the writing and editing process, contributing to the bottom-up learning process among troops at various levels.[38]

The manual's popularity resulted in the University of Chicago Press publishing it in the spring of 2007 with a new introduction by Sarah Sewall, the director for the Carr Center for Human Rights at Harvard's Kennedy School and a Clinton-era former deputy assistant secretary of defense for humanitarian operations and peacekeeping. In this thought-provoking essay, the human rights expert discussed how much of a radical departure this new doctrine is for the conventional military: "The counterinsurgency field manual challenges all that is holy about the American way of war... Those who fail to see the manual as radical probably don't understand it, or at least don't understand what it is up against."[39]

To help him implement this "radical" approach during the Surge, General Petraeus deployed with an inner circle of advisers comprised of key military and civilian COIN experts from government as well as academia.[40] This group included Ambassador Robert Ford from the State Department; Derek Harvey, a civilian intelligence analyst and former Army officer; Dr. David Kilcullen, a civilian counterterrorism official and former Australian Army infantry officer; Colonel Peter Mansoor, a former brigade commander and military historian; Colonel H. R. McMaster, whose record as a brigade commander in Tal Afar, Iraq, the previous year was considered a model for counterinsurgency in practice; and Colonel Mike Meese, a professor at West Point who served with Petraeus in Mosul.[41] In addition, scholars such as Dr. Steve Biddle from the Council on Foreign Relations and Toby Dodge from King's College were also invited to participate in the initial assessment. Many of these scholars,

---

38. John Nagl, foreword to University of Chicago Press edition, *The U.S. Army/Marine Corps Counterinsurgency Field Manual* (Chicago: University of Chicago Press, 2007).

39. Sarah Sewall, introduction to University of Chicago Press edition, *The U.S. Army/Marine Corps Counterinsurgency Field Manual* (Chicago: University of Chicago Press, 2007).

40. Thomas Ricks, "Officers with PhDs Advising War Effort," *Washington Post*, February 5, 2007.

41. Packer, "The Lesson of Tal Afar."

Learning to Surge in Iraq • 179

soldiers, and civilian practitioners had also helped craft the COIN manual the previous year and shared General Petraeus's frustration with the Iraq strategy to date. Together they turned a 50-page idea from a think tank into a real operational strategy, including new tactics and commander's guidance.[42]

The Surge strategy meant a significant shift in focus for coalition troops on the ground. As the letter titled "Counterinsurgency Guidance" issued by the American ground commander Lieutenant General Raymond Odierno (and also published and distributed in Arabic) made perfectly clear, their primary role would now be to "secure the people where they sleep."[43] Instead of retreating into a smaller and smaller number of large forward operating bases (FOBs) in order to reduce the U.S. footprint and turn the task of security over to inexperienced Iraqi security forces, U.S. troops would now be moved off of these "uber-FOBs" (as they had come to be called) into numerous smaller outposts deployed among the residential population all over Baghdad. General Petraeus's chief counterinsurgency advisor, Dr. David Kilcullen, explained the rationale for the shift in strategy in a popular online blog post in January 2007.

> The new strategy reflects counterinsurgency best practice as demonstrated over dozens of campaigns in the last several decades: enemy-centric approaches that focus on the enemy, assuming that killing insurgents is the key task, rarely succeed. Population-centric approaches, that center on protecting local people and gaining their support, succeed more often. The extra forces are needed because a residential, population-centric strategy demands enough troops per city block to provide real and immediate security.[44]

This "best practice" was also reflected clearly, and not incidentally, in the counterinsurgency field manual (FM 3-24) discussed above. Thus the choice of General Petraeus to lead the Surge was significant, as he represented a group of military officers and academics who had been pushing for a new, more counterinsurgency-focused approach to the

---

42. Linda Robinson, *Tell Me How This Ends: General David Petraeus and the Search for a Way Out of Iraq* (Public Affairs, 2008).

43. Lieutenant General Ray Odierno, Commander, Multi-National Corps, Iraq, "Counterinsurgency Guidance" (letter issued to Coalition troops in Baghdad, Multinational Corps Iraq [MNC-I], June 16, 2007).

44. David Kilcullen, "Don't Confuse the 'Surge' with the Strategy," *Small Wars Journal* blog, January 19, 2007, http://smallwarsjournal.com/blog/2007/01/dont-confuse-the-surge-with-th/.

conflict in Iraq.[45] The question was, would it work, or was it simply too late politically and otherwise to make these changes? Had time run out on the "Washington clock," or would Congress and the American people give Petraeus a chance?

Assessing the Surge

The primary objective of the Surge was to get the skyrocketing violence under control so that the Iraqis could focus on the political process. "Violence" in this context meant not only attacks on coalition forces, which heretofore had been the primary metric routinely tracked by U.S. forces, but also violence among the Iraqi population, which had reached alarming rates and included gruesome acts of slaughter and torture by the summer of 2006. The fact that U.S. commanders had begun to recognize that success in a complex counterinsurgency environment had to be measured with a focus on the local civilian population, not just on traditional combatants, is one of many indications that a significant degree of organizational learning had occurred since 2003.

The security situation, as measured by civilian and coalition military incidents, improved markedly during the first year of the Surge, while the political process remained stagnant. By late spring 2008, civilian deaths had dropped dramatically to pre-2006 levels, U.S. casualties were at their lowest rate since the war began (although there were months with temporary spikes), and the influence of Al Qaeda in Iraq (AQI) had been greatly reduced.[46] As the debate over the wisdom of the Surge subsided, pundits and experts argued over whether the undeniable improvements in security were actually a result of the Surge's new tactics and increased numbers of security forces or were due to other developments. Could the drop in violence be due to the fact that ethnic cleansing and sectarian violence had possibly run their course in the year prior to the Surge? Likewise, could the coalition actually take credit for the so-called Tribal Awakening in which Sunni tribes turned against Al Qaeda in Iraq, given that this turn had begun before the Surge?[47]

While it is difficult to determine an absolute cause-and-effect relationship between the Surge and the drop in violence, a strong case can

---

45. Spencer Ackerman, "The Colonels and 'the Matrix,'" *Washington Independent*, July 28, 2008, first article in the series The Rise of the Counterinsurgents.

46. Colin Kahl, "Shaping the Iraq Inheritance" (Washington, DC: Center for New American Security, June 2008); O'Hanlon and Campbell, "The Iraq Index."

47. Celeste Ward, "Challenging the Military's New Mantra," *Washington Post*, May 17, 2009.

be made that the success of the Awakening is in part a function of the new strategic environment created by the Surge—which was in turn enabled by a significantly seasoned, better-trained and better-educated military than had begun the campaign four years earlier. The new tactics implemented by better-prepared and experienced commanders allowed for unprecedented cooperation among coalition forces and Sunni tribal leaders. U.S. commanders were encouraged to interact and make deals with local leaders, and, remarkably, former insurgents who had grown tired of AQI's growing influence in their regions were open to partnering with the coalition to run AQI out. In a comprehensive report prepared by the Center for a New American Security, Dr. Colin Kahl describes how, as a result of the new approach applied in the Surge, commanders were able to capitalize on the emerging Awakening.

> Nimble U.S. commanders effectively exploited the growing wedge between Sunni tribes and AQI to forge cooperative arrangements, and the tribes responded by providing thousands of men to serve in auxiliary security forces . . . The result was a dramatic reduction in violence in Anbar, once the hotbed of the Sunni insurgency . . . Although the Surge did not spark the Awakening, the new American approach did help it spread outward from Anbar. Even here, the real cause was not the additional troops per se, but rather a change in strategy associated with the Surge.[48]

The American military's response to the 2007 Sunni Awakening in al-Anbar province is a stark contrast to the lack of engagement with locals that was the norm for most of the invading and occupying U.S. military of 2003. It required not only improved understanding of COIN principles but also a sophisticated awareness of the cultural environment. Previous events of major cultural significance to the Iraqis, such as the bombing of the Golden Shrine in Samarra in February 2006, which ignited a tidal wave of ethnic violence, had gone virtually unnoticed by the coalition. Whereas it took five months for the coalition to respond to the Samarra bombing, exploitation of the Sunni uprising and the application of the "Anbar model" elsewhere in Iraq were relatively swift.[49] This cognitive and strategic shift reflects a high degree of learning and adapting as well as the enlightened leadership of General Petraeus and his team. In sum, exploitation of the Awakening would not have been possi-

---

48. Kahl, "Shaping the Iraq Inheritance."
49. David Kilcullen, "Dinosaurs versus Mammals: Insurgent and Counterinsurgent Adaptation in Iraq, 2007" (presentation to RAND Insurgency Board, Washington, DC, May 8, 2008).

ble without the Surge, but the Surge would not have been possible had the American military not already been on a learning curve from the previous four years in theater.

Organizational Learning: The Foundation for the Surge

Implementation of Petraeus's Surge strategy, along with the sometimes counterintuitive change in tactics it demanded, would have been much more difficult had the force not learned collectively since 2003 and had a better-developed culture for contemporaneous adaptation. For example, the commander's guidance issued to shift the strategic and tactical focus to protecting the people versus primarily targeting enemy fighters also urged troops to "get out [of your armored vehicle] and walk."[50] This guidance was based on classic COIN theory that emphasizes personal interaction with locals as well as on field research by Petraeus's team that determined that troops were more vulnerable to roadside bomb attacks when they were in vehicles than they would be to sniper attacks were they to dismount.[51]

To ensure the implementation of such new guidance, the MNF-I leveraged "accelerant tools," which included a number of in-theater adaptations, some of which had developed over time in theater prior to the Surge.[52] These included the following:

- Enhanced in-theater training. A "graduate COIN course" was provided by the Counterinsurgency Academy in Taji, Iraq, set up by General Casey before the Surge. This school ensured that incoming units were indoctrinated on the latest tactics, techniques, procedures, and lessons learned that filtered up from across Iraq. Similar courses were later set up in the embassy to educate incoming civilians on principles of counterinsurgency and new Surge principles, to ensure that civilians and military were on the same sheet of music.

- Field "coaches." Petraeus's senior COIN advisor, David Kilcullen, acted as an on-field "coach," traveling from the commander's headquarters to the various units to help the field commanders interpret the latest guidance. An experienced Australian infantry officer himself, Dr. Kilcullen was also able to accompany units on actual missions, where he provided technical advice, collected data, and conducted research on what was and was not working. In this role, he facilitated a feedback loop to headquarters,

---

50. Odierno, "Counterinsurgency Guidance."
51. Kilcullen, "Dinosaurs versus Mammals."
52. Ibid.; Kahl, "Shaping the Iraq Inheritance."

which allowed fine-tuning to strategy as required. Most important for contemporaneous adaptation, Dr. Kilcullen cultivated lateral communication among units by linking commanders to each other via e-mail and fostering communities of practice for information sharing. Dr. Kilcullen's efforts were enhanced by coaching teams that were deployed across the theater to provide units with more tactical-level advice.[53]

- Formal and informal doctrine. The widespread dissemination of the new COIN field manual, FM 3-24, provided an intellectual foundation for new guidance (i.e., Multi-National Corps–Iraq [MNC-I] commander Lieutenant General Odierno's "Counterinsurgency Guidance"). Informal doctrine, such as the *Counterinsurgency Reader*, which was a compilation of key articles published in the Army's popular journal *Military Review*, was issued in the summer of 2006. Both FM 3-24 and *Military Review* were assigned in all the courses and available online as well as distributed in hard copy.[54]

- Mature communities of practice. Online communities of practice, as discussed in previous chapters, had become common methods of information sharing among commanders at all levels. In addition to the popular CompanyCommand.com website, military and civilians in theater regularly accessed Army-sponsored sites such as PlatoonLeader.com, West Point's counterterrorism site, and the Combined Arms Center blog. Another popular nonmilitary site was the online Small Wars Journal blog, a site run by two retired marines, Dave Dilegge and Bill Nagle. Dilegge describes their audience of nearly 400,000 as "niche, . . . but one that spans all ranks and increasingly non-DoD and Non-US. Moreover, many journalists and academia use the page for basic and more advanced research."[55] The site hosts a number of well-respected theorists and practitioners who regularly post articles and comments to the site. Petraeus's COIN advisor also blogged on this site during the Surge to help people better understand the strategy, and readership spiked during those months. According to Dilegge, as of summer 2008, these posts from the field "remain as the most popular and the most cited," reflecting a desire to better understand the emerging lessons and strategy. (Small Wars Journal is working to publish the collection as a book.)

- Enhanced cultural awareness. Training to enhance cultural awareness had been added at all levels of the system. A shift in focus toward understanding the "human" as well as the traditional military "physical" terrain

---

53. Kilcullen, "Dinosaurs versus Mammals."
54. All volumes of *Military Review* are available online at http://usacac.army.mil/CAC2/MilitaryReview/.
55. Dave Dilegge, personal correspondence with the author, 2008, Washington, DC.

improved intelligence collection and interaction with locals and was a prerequisite to implementing demographic targeting of insurgent networks.[56] As one Army War College researcher observed, "the recent focus on cultural knowledge in counterinsurgency operations and tactics is a welcome development insofar as it has allowed field commanders in Iraq and Afghanistan to radically reassess the failed operations and tactics in counterinsurgency in both these places."[57]

These in-theater "accelerants" enabled the commander's intent to be implemented more uniformly across time and space and also allowed for rapid adaptation as required. The development and implementation of these "accelerants" and the military successes of the Surge as a whole reflected the enhanced post-Vietnam learning capacity of the U.S. military. Key adaptations and lessons learned by individual commanders, such as methods for interacting with local leaders, patrolling techniques, and network analysis, were shared online, collectively analyzed in theater and back at the stateside schoolhouses, and incorporated into the countrywide strategy.

The most obvious personification of this learning curve was General Petraeus's ground commander, Lieutenant General Ray Odierno, who issued the enlightened "Counterinsurgency Guidance" as MNC-I commander in 2007. Odierno had been widely criticized for his heavy-handed approach on his previous deployment as the Fourth Infantry Division commander. Far from the progressive MNC-I commander he had become by 2007, his leadership as division commander in 2004 had been, according to author/journalist Tom Ricks, "almost the opposite of Petraeus's 101st Airborne."[58] By 2007, however, Odierno was in lockstep with General Petraeus, reflecting the cycle of learning of the institution as a whole.

In sum, by all accounts, the troops that Petraeus commanded were better prepared to comprehend and operationalize his guidance than they would have been in 2003. Still, despite the security gains made during the first 18 months of the Surge compared to prewar levels of violence and stability, Iraq was still struggling. By summer 2008, with the

---

56. "Demographic targeting" refers to a method of tracking urban insurgents based on their propensity to return to their rural hometowns. Tracking movements to and from rural areas vis-à-vis knowledge of kinship lines helps identify insurgents and break up networks. Kilcullen, "Dinosaurs versus Mammals."

57. Sheila Miyoshi Jager, "On the Uses of Cultural Knowledge," Strategic Studies Institute Published Monograph (Carlisle, PA: U.S. Army War College, 2007), http://www.au.af.mil/au/awc/awcgate/ssi/jager_cultural_knowledge.pdf.

58. Ricks, *Fiasco*.

U.S. presidential election in full swing, the political gains promised as part of the Surge had not materialized, Iraqi prime minister Nuri al-Maliki was calling for a timetable for withdrawal, and the U.S. military was close to the breaking point from sustained rotations of troops and matériel. This state of affairs suggests that bottom-up military learning is a necessary but insufficient condition for success in stabilization and reconstruction operations.

The Limits of the Learning System: The Challenge of the Four-Block War

The key problem with the military's learning system in meeting today's challenges is that it is observationally and tactically oriented and thus does not work well beyond the environment of the so-called three-block war. The three-block concept was best articulated by Marine Corps commandant General Charles Krulak in 1997.

> In one moment in time, our service members will be feeding and clothing displaced refugees, providing humanitarian assistance. In the next moment, they will be holding two warring tribes apart—conducting peacekeeping operations—and, finally, they will be fighting a highly lethal mid-intensity battle—all on the same day . . . all within three city blocks.[59]

Krulak's "three-block war" resonated with a generation of U.S. warfighters who had increasingly found themselves conducting complex and frustrating missions from Somalia to Kosovo that, in the words of Major General Steve Arnold, "may not be war, but sure as hell ain't peace."[60] Although the U.S. Army was at the same time developing a similar (if less pithy) doctrine called "full-spectrum operations," the "three-block war" image was simple and to the point. It gave troops a conceptual "handrail" and had a profound impact on the U.S. military's operational and tactical mind-set. U.S. forces would no longer deploy with an expectation for fighting *either* a war *or* an "other-than-war" operation. Rather, they would deploy with the very real expectation that they would need to conduct three different, but interrelated, types of operations—*simultaneously.*

---

59. Charles C. Krulak, "The Three Block War: Fighting in Urban Areas" (speech presented at National Press Club, Washington, DC), *Vital Speeches of the Day*, December 15, 1997, 139–41.

60. Steven L. Arnold, "Somalia—an Operation Other than War," *Military Review* (December 1993).

What the concept of three-block war leaves out, however, are the more complex reconstruction, governance, rule-of-law, and economic development types of tasks required to counter insurgents and other violent spoilers, win the hearts and minds of the population, and restore a broken country to a functioning society—that is, those tasks required to achieve the strategic objectives of the coalition mission in Iraq. Such tasks constitute a fourth block of operations and are the area in which the U.S. military—indeed, the U.S. government—still struggles today.

Although the military's learning system enables troops to learn contemporaneously better than ever before, integrated four-block tasks require more than tactical-level, observationally oriented learning. The dynamic learning process described in this book enables ground-level lessons on what does and does not work to be transferred quickly throughout the organization and especially into the immediate training cycle. These lessons are captured through the immediate observation of events by "collection teams" of observers who attempt to identify what is working and what is not. Their observations are then rapidly disseminated throughout the system via myriad media, including Internet sites, handbooks, and training scenarios.

But what happens when the "lesson" on the ground is not instantaneously obvious—especially to a "collector" that has been schooled in traditional military arts? For example, how does a commander or observer know that the microfinance program she initiated in an Iraqi province has worked or that a brigade's choice to build a school instead of a town hall, mosque, or marketplace was the right decision? Moreover, as development experts will assert, such projects often have second- and third-order effects and unintended consequences that may not become evident for years.[61] How can the military's learning system, which is generally premised on instantaneous observation, account for this?

Although the military's learning in relation to three-block war has improved greatly, the system is still contingent on understanding what "success" or "victory" is through the process of simple tactical-level observation. At the ground level, this is much easier to do for three-block tasks, such as running a checkpoint or convoy, setting up a refugee site, or cordoning and searching a house, than for reconstruction or nation-building tasks, such as resurrecting a government, kick-starting an economy, and otherwise resolving the underlying conflict. When such state-building tasks are tied to long-term (and often ambiguous) strategic objectives (i.e., "promote democracy"), capturing lessons from the field becomes a

---

61. Hegland, "National Security—Why Civilians Instead of Soldiers?"

much more complex, more political, and slower endeavor. Conducting four-block tasks—and collecting lessons on these missions—requires theoretical knowledge on subjects as diverse and complex as democratization, economic development, city planning, and the rule of law.

The biggest challenge to injecting theoretical state-building knowledge into the lessons-learned system is that the community of practitioners and experts simply do not agree on how to do it. Although doctrine on COIN and stability operations has been updated to include four-block elements such as economic development, governance, and rule of law, the doctrine is little more than "PowerPoint deep," reflecting a lack of theoretical consensus. Indeed, after four years in Iraq and Afghanistan and a decade of experience conducting similar missions in the 1990s, officials in the field are still arguing over the right approach and, in many cases, using completely different sets of theories.

Is the right approach about privatization and economic "shock therapy," as many economists claim, or providing jobs first, as Lieutenant General Chiarelli claims? Should we shut down state-run industries in the name of privatization, as advocated by Paul Bremer, or allow them to operate because they provide jobs, as claimed by senior defense official Paul Brinkley? What about crime and corruption? Should we try to marginalize organized crime cells, or should we pull these power brokers in closer and try to co-opt them? How should we prioritize our use of CERP (Commanders Emergency Response Program) funds? Should we start a microfinance program or build a medical clinic or school? Should we make locals pay for electricity and other essential services or offer it free of charge to jump-start economic life? Should we "brand" our presence and take credit for doing good things so the media can tell good tales and sustain support of the American people and the international community, or should we take an indirect route and allow local actors to take the credit? What about enhancing governance? How should we build the "rule of law" at the local and national levels? And should these activities wait until after we "establish" security—or is security itself contingent on getting the economy and the government back up and running and getting angry young men off the street and kids back in school (so parents can go to work)? Finally, how should we measure progress and determine that we are not making things worse in the long term in our quest to solve short-term problems? In short, as long as the "experts" are still debating on foundational theories in an environment where success is not immediately obvious given the myriad second- and third-order effects inherent in the state-building tasks, collecting lessons in this fourth block will continue to be a challenge.

We must also keep in mind that the military's learning system is designed for the military—not for the military operating with other agencies. Despite the fact that, as discussed previously, civilian agencies lack capacity to operate in large numbers, they are still increasingly part of the stability operations landscape. A robust stability operations lessons-learned system will need to incorporate these civilian actors, and the agencies in which they work will need to develop more of a learning culture.

The struggle to learn in the complex interagency environment of the four-block war should not, however, diminish the fact that the U.S. military has *learned to learn* in the three-block war, where they are the primary actor, and that this learning has enabled the modest successes in reining in violence during the Surge. Although there is still more work to be done to develop a comprehensive learning process across the U.S. government as a whole, understanding how the military was able to make this shift from the 1970s to the present provides a valuable lesson in organizational change.

Making Sense of Iraq and Beyond

The emerging narrative on the American military's experience in Iraq is that a military that had wholly and intentionally forgotten about its past experiences with counterinsurgency invaded Iraq unprepared to meet the challenges it faced. This was due to the military's cultural predisposition to "big war" and its aversion to counterinsurgency or other so-called irregular operations. It was widely assumed at the time that this military was incapable of learning and adapting—as it had proven in its experience in Vietnam. In a widely circulated and highly critical article in the American Army journal *Military Review,* British brigadier Nigel R. F. Alywin-Foster compared the culture of the modern U.S. Army with Nagl's Vietnam-era institution and observed,

> There is plenty of evidence, from Nagl by implication, and from other sources more directly, that this uncompromising focus on conventional warfighting, and concomitant aversion to other roles, have persisted to the present day, or at least until very recently, and were instrumental in shaping the Army's approach to OIF in 2003 and 2004.[62]

---

62. Brigadier Nigel Aylwin-Foster, British Army, "Changing the Army for Counterinsurgency Operations," *Military Review* (November–December 2005). See also a refutation of this article by Janine Davidson ("Aylwin-Foster Misunderstands Nagl's Army," *Military Review* [January–February 2006]), in which the author observes that the Army has changed and that it is on a learning curve in Iraq.

To the surprise of many observers (especially those who had read John Nagl's book on the Army's failure to adapt in Vietnam[63] and had accepted the criticism from the British brigadier), this assumption was proved false. Somehow, the narrative goes, this military managed to adapt in Iraq, making Petraeus's Surge strategy possible.

Contrary to emerging conventional wisdom, this adaptation was not due solely to the brilliant leadership of a handful of enlightened generals and colonels—such as Petraeus, Mattis, Chiarelli, and their teams—who pushed COIN doctrine and strategy on an unwilling military establishment. While these leaders did play critical roles in the institutional adaptation that took place, they were not a product of, nor were they leaders in, the Vietnam-era Army. The military that crossed the line in Baghdad in 2003 was simply not the same military that failed to learn in Vietnam, and this hard-won shift in organizational culture is the critical factor that enabled the military to adapt.

As this book has demonstrated, through visionary leadership and institutional changes made in the decades following Vietnam, this military—especially the Army—had become more of a "learning institution." Through institutional changes, the development of an after-action process for "getting to the ground truth," and a tactical, training-oriented lessons-learned process, the military's culture had shifted. These processes had taught a generation of leaders and their troops how to learn. This generation had developed a "get me the answers" culture and was predisposed to question, evaluate, share information and new ideas, and adapt on the fly.

More substantively, this generation's experience in the 1990s with so-called MOOTW (military operations other than war) had informed doctrinal changes and training (though education was still lagging) and had a profound impact on a significant population of military officers who had served in the complex and chaotic stability operations in Somalia, Haiti, and the Balkans. Many of the more-progressive initiatives implemented by commanders, such as First Cavalry Division commander Lieutenant General Peter Chiarelli's use of SWETI (sewage, water, education, trash, and information) lines of operation, had been developed on the ground in the Balkans and written into training scenarios at JRTC. While the Army's Peacekeeping Institute was instrumental in getting this experience written into new "stability operations" doctrine, the Special Forces had effectively kept the intellectual candle burning on the devel-

---

63. John A. Nagl, *Counterinsurgency Lessons from Malaya and Vietnam: Learning to Eat Soup with a Knife* (London: Praeger, 2002).

opment of thought, theory, and doctrine for counterinsurgency. All of this experience was leveraged when it came time to write a new COIN field manual for the rest of the military.

Still, as the case of Iraq also demonstrates, military learning, no matter how comprehensive or how swift, is necessary but not always sufficient for strategic success in complex stabilization and reconstruction operations or counterinsurgency. When the strategic objectives require success in the "fourth block," a realm that often involves a number of nonmilitary actors and an environment in which "lessons" are slow to emerge, the military's system of rapid tactical learning has limited utility. Likewise, as has been the case throughout history, individuals matter. Leaders can set command environments that permit, promote, or prevent learning to occur. Unenlightened civilian or military leadership can make strategic or policy decisions that undermine a bottom-up learning trajectory or otherwise complicate conditions for those on the ground. This was certainly the case with the myriad poor planning assumptions between CENTCOM and the Pentagon and, again, with Paul Bremer's CPA.

In sum, the lessons-learned system currently in use must be modified to accommodate these issues. More research is needed to determine what processes would work in the four-block interagency environment. Analyses of on-the-ground observations must be informed by more in-depth research, including trend analysis on historical cases and basic theories from myriad academic disciplines, in order to make sense of what is being observed. Whether or not this added step will allow as rapid a turn cycle as the current system utilizes and as the current generation of military professionals has come to expect remains to be seen.

Conclusion: Learning Theory and Military Change in the 21st Century

I am tempted to say that whatever doctrine the armed forces are working on now, they have got it wrong. I am also tempted to declare that it does not matter . . . What does matter is their ability to get it right quickly, when the moment arrives.
—SIR MICHAEL HOWARD[1]

Learning implies drawing the proper lessons from the events and spreading the experience among others. All this cannot be left to chance and personal initiative; it must be organized carefully and deliberately.
—DAVID GALULA[2]

For over 200 years, the U.S. military has conducted operations other than "major war," including nation building, counterinsurgency, stabilization and reconstruction, and peacekeeping. Rarely, however, has the institution translated this experience into institutional learning, leaving each generation of military commanders with only personal experience, ingenuity, and initiative to accomplish the mission. Yet despite purported cultural resistance to conducting such missions, an examination of the post–Cold War U.S. military indicates a subtle shift in this trend. New doctrine, new training scenarios, and, lately, new educational curricula based on the past 15 years of operational experience reflect an effort toward institutional learning. This shift is a result of

---

1. Michael Howard, "Military Science in an Age of Peace," *RUSI Quarterly* 119, no. 1 (1973). Quoted in Gordon Sullivan and Michael V. Harper, *Hope Is Not a Method: What Business Leaders Can Learn from America's Army* (New York: Random House, 1996).
2. David Galula, *Counterinsurgency Warfare: Theory and Practice,* 2nd ed. (London: Praeger Security International, 2006). The first edition was published in 1964.

post-Vietnam structures and processes that were actively designed to capture and disseminate experiential knowledge.

In his comparison of the U.S. Army in Vietnam and the British Army in Malaya, John Nagl emphasized organizational culture as the primary obstacle to or catalyst for contemporaneous institutional learning and military change. "It is," he writes, "the organizational culture of the military institution that determines whether innovation succeeds or fails."[3] Using organizational learning theory, Nagl concluded, "The organizational culture—the 'persistent, patterned way of thinking about the central tasks of and human relationships within an organization'—played a key role in allowing an organization to create a consensus either in favor of or in opposition to proposals for change."[4] Indeed, the research presented here also highlights organizational culture as a significant obstacle to military change. It is not, however, insurmountable or completely determinate.

In contrast to theories of military change that focus on organizational culture and external political intervention as the most significant obstacles to and catalysts for change, the research presented here demonstrates how internal institutional structures and processes can *prevent*, *promote*, or *permit* military change through learning. Whereas previous attempts at disseminating and institutionalizing experiential learning were often stymied by the military's cultural preference for "big war" and its structural inabilities to transfer knowledge to the next generation, the case of the post–Cold War Army suggests that organizational systems designed to actively capture and disseminate new experiential lessons can act as a counterweight to cultural resistance and organizational change.

In contrast to the Vietnam-era Army studied by Nagl and others, today's Army presents a case for how formal institutional structures and processes that are consciously designed to generate contemporaneous institutional learning can act as a powerful counterweight to entrenched organizational culture. With organizational structures and processes in place that are designed to permit and even promote the transfer of experiential knowledge, traditional cultural preferences for sustaining an organization's "essence" can be effectively challenged by new operational experience. This is exactly what the post-Vietnam Army's learning system was designed to do and how it operated throughout the 1990s at

---

3. John A. Nagl, *Counterinsurgency Lessons from Malaya and Vietnam: Learning to Eat Soup with a Knife* (London: Praeger, 2002).

4. Ibid.

the tactical level. Although the system is still susceptible to interference by powerful and savvy political actors, the modern Army's internal propensity for learning in more recent years, compared to previous eras, has had a demonstrated effect on its capacity to learn.

Institutional Amnesia or Lessons Learned? A Mixed History

Conducting operations other than war is nothing new for the U.S. military. Unfortunately, military doctrine, education, and training have rarely prepared troops well for these tasks, and the institution has not managed to incorporate lessons from these operations into doctrine, education, and training for the next generation. The perennial mismatch between operational experience and institutional knowledge is a reflection of both military and civilian cultural preferences. Culturally, American society has resisted the use of its military forces for imperial purposes. U.S. military professionals have reflected these preferences by limiting their perceived role to "fighting and winning the nation's wars," rather than internalizing a broader conceptualization that would include more limited uses of force. Popular theories of "big war" as prescribed by European theorists such as Jomini and Clausewitz have reinforced this preference. For commanders throughout history who have been sent to accomplish tasks other than "major war," however, these two cultural predispositions have resulted in strategic and political ambiguity over mission objectives and military roles, complicating commanders' abilities to operate.

Throughout history, commanders have coped with this ambiguity by relying on a "handrail" of personal experience and the experience of others. Failing the availability of either of these, they have coped through trial and error on the job. This has resulted in inconsistent approaches across time and space within the same operation, depending on the personality, experience, and knowledge of the commanders on the ground. Occasionally, veterans of these "nontraditional" missions have attempted to overcome this cultural and organizational inertia by sharing their experiences with each other and with the next generation. They have attempted this through informal networks and military journals and, when allowed or positioned to do so, through changing formal doctrine, education, and training. These attempts at institutional learning have been moderately successful when high-level leaders have offered their support and when formal structures and processes existed through which their ideas could be disseminated.

Military Learning: What Has Worked and How

Amid the perennial institutional amnesia in the U.S. military, there have been a few minor victories by military veterans of irregular operations. Their examples are worthy of examination in order to identify the mechanisms that permit or promote learning and change. For example, the authors of the *Small Wars Manual* (published in 1940) were able to transfer the lessons learned from two decades of collective personal experience in the Banana Wars to educational curricula and formal doctrine. Key factors in their success were (1) institutional mechanisms in the form of the Quantico School and the official *Marine Corps Gazette* and (2) the fact that the general in charge of the Quantico School supported their efforts. In doing so, this commander countered the Marine Corps commandant's preference for amphibious warfare studies and the prevailing cultural resistance to small wars and demonstrated the importance of leadership in promoting institutional learning.

This laudable attempt to institutionalize collective contemporaneous learning was not sustained, however. By 1944, the *Small Wars Manual* was no longer incorporated into formal Marine Corps education, as all efforts were focused on the problem of amphibious warfare during World War II. By 1962, when a new manual for counterinsurgency was written, the 1940 *Small Wars Manual* had been long forgotten. Still, although the cultural preference for amphibious warfare remained prominent in the Corps for decades, it is important to recognize how experiential knowledge about small wars had been formally captured, if not formally institutionalized. This was more than the Army had accomplished following their many small wars experiences of the 19th century.

The Army experience in the 19th century was characterized by a continuous series of small wars and stability operations. Similar cultural aversion—especially to colonization and occupation—sent mixed messages to commanders operating in the American South, the Philippines, and Cuba. Not until the interwar years prior to World War II did new processes for war planning, along with a newfound emphasis on professional military education and history, allow one group of committed Army veterans to break the cultural barriers to studying occupation duty. Using these new processes, they were able to study and publish their World War I occupation experiences as doctrine and to create schools and educational curricula for civil affairs and military governance. Although their efforts were moderately stymied by military and civilian leaders opposed to the use of the U.S. military as "occupiers," without their efforts—and the organizational structures that supported those efforts—the post–World War II occupation would have been far worse off. This episode demonstrates the importance of organizational systems (i.e., the war-planning process, the history office, and the war colleges) through which innovators can operate by planting ideas as seeds so change can be initiated. Similar processes also worked for a generation of peacekeepers in the 1990s who benefited from a new training system, including a formalized lessons-learned system and the small Peacekeeping Institute at the Army War College.

Doctrine, Education, and the Training Revolution

Following the Vietnam War, the U.S. Army developed a new system for capturing experiential knowledge from both training and real-world operations. The post-Vietnam system reflects a conscious effort by this generation of military leaders to drive change and thus reflects the improved peacetime, top-down method of institutional learning. The three coordinated tiers to this system were doctrine, education, and especially training. The new doctrine that these leaders created for warfighting, called the "AirLand Battle Doctrine," was a deliberate attempt to return the Army to its primary focus of preparing to fight a large European land war against the Soviets. The doctrine was "validated" though experimentation, injected into educational curricula throughout the Army, and practiced at the new training center in Fort Irwin, California (fig. 4). Thus education and training were in sync as they were driven by a top-down injection of new doctrine derived from theory. Dissemination of the new doctrine was further accelerated through the School of Advanced Military Studies (SAMS), from which the elite field-grade officer graduates were sent directly to high-level planning staffs to practice their new operational art and to spread the new message up the chain.

The backbone of this post-Vietnam system was the combat training centers (CTCs), which spawned the Center for Army Lessons Learned (CALL). This system indoctrinated a generation into a new way not only of warfighting (AirLand Battle Doctrine) but also of "learning through doing," through the use of the process of after-action review and large force-on-force exercises. Although the system was designed to reinforce the big-war paradigm, real-world lessons from operations in Somalia, Haiti, and the Balkans throughout the 1990s were also able to travel along these same institutional rails. This created a mismatch between what was being taught in the formal classrooms and what was being practiced in the training centers and experienced in real life. Soldiers were learning through doing but were being educated in the old doctrine,

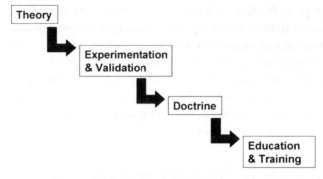

Fig. 4. Top-down, theory-driven learning system

with an emphasis on the paradigm of the AirLand Battle Doctrine. During this time, doctrine was steadily revised to reflect some of this operational experience. Although this doctrine was immediately filtered through the training centers via new scenarios and made available via the Internet and through printing offices to anyone wishing to self-educate, it was not emphasized in professional military education (fig. 5).

Training. As figure 5 reflects, the learning system reacted to the operational experiences of this generation by steadily compiling lessons learned from each operation, feeding these lessons into the training centers, and exposing troops to this knowledge during their mission rehearsal exercises (MREs) at the Joint Readiness Training Center (JRTC) and the Combat Maneuver Training Center (CMTC), where units were sent to prepare for deployment to Haiti and the Balkans. This dissemination system was complemented by the age-old practice of tapping into informal networks. Today, these networks are facilitated by the Internet. What is different than in the past is the conscious effort by the institution to access this experiential learning and to accelerate informal networking.

Although the National Training Center (NTC) in Fort Irwin, California, remained focused on large tank-on-tank maneuvers until 2003, both the JRTC in Louisiana and the CMTC in Germany adjusted their scenarios significantly to prepare units for their deployments in the Balkans. These scenarios were continually updated based on lessons learned in the field, as described in chapter 4. Units who, for scheduling reasons, could not attend a formal training center exercise prior to deployment mimicked these scenarios at their home stations to prepare their troops. Today, tactical lessons from Iraq and Afghanistan are fed directly to these

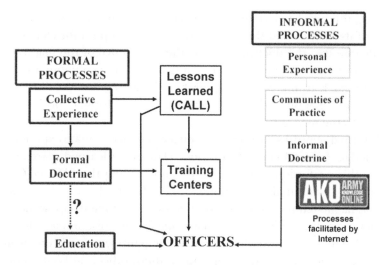

Fig. 5. Post–Cold War bottom-up experience-driven learning system

centers, where scenarios are adjusted immediately to reflect the latest challenges in the field.

Doctrine. Throughout the 1990s, changes in doctrine occurred at a steady pace. These changes were initiated by midlevel officers who were veterans of Somalia, Haiti, and the Balkans, and the changes thus reflected their experiences. Still, powerful preferences for traditional warfare remained throughout the institution—especially in the senior leadership. This generational split was reflected in some of the language published in these manuals, as many of the MOOTW-oriented manuals written at this time contained caveats added by leadership reflecting institutional ambiguity over the relative importance of these types of missions. By the mid-1990s, when the mismatch between operations, on the one hand, and doctrine and education, on the other, was greatest, attempted updates to manuals such as the 1996 drafts of FM 100-5 *Operations* (which attempted to introduce "full-spectrum operations") and FM 100-20 *Stability and Support Operations* were stymied by internal disagreement and high-level resistance. It took five and seven more years, respectively, for these manuals to be approved for publication. This shift also corresponds to a new generation of leaders assuming positions of authority in the process of doctrine approval.

In contrast to the AirLand Battle Doctrine of the previous era, which was based on theory and used as a top-down catalyst to drive institutional change, this new batch of doctrine was based on the operational experience of its authors and thus reflected a bottom-up push for change. To the degree that the doctrine reflects the experience of the majority of the institution, it can be seen as a trailing indicator of contemporaneous institutional learning. Sustaining this knowledge requires the dissemination of this experienced-based doctrine, through formal and informal means, to the next generations.

Informal networks have always been a way to disseminate experiential knowledge. General James Mattis once claimed, "I learned more about life in this profession at happy hours and reading the [*Marine Corps*] *Gazette* than I did in all my training/PME."[5] Lately, the use of the Internet has facilitated the use of informal networks among troops while also accelerating the military's formal capacity to disseminate lessons learned. The fact that websites such as CompanyCommand.com, which was independently started by a group of junior officers at West Point, is fully supported (and copied) by the Army leadership reflects how far Army leaders have come in understanding organizational learning theory.[6] Meanwhile, the military makes extensive use of technology to post lessons learned on official sites such as the site for the Center for Army Lessons Learned, Joint Knowledge Online, and many other unit-specific sites. Together, these systems facilitate the acceleration of the contemporaneous learning loop at the tactical level. Still, such systems are more passive than active. They *permit,* rather than *promote,* the dissemination of bottom-up knowledge. For a more complete learning cycle, formal education must be integrated with these systems.

Education. The weak link in the post–Cold War learning system was professional military education (PME). Throughout the 1990s, while the training centers and CALL were capturing and disseminating new lessons from operations in Somalia, Haiti, and the Balkans, PME continued to limit study of these topics to a small percentage of the overall curricula. Thus, even though new doctrine was written for MOOTW missions throughout the 1990s, little time was allotted in these schools for it

---

5. James Mattis, Lieutenant General, USMC, personal interview by author, June 15, 2005, Quantico, VA.

6. General Kevin Byrnes, USA, Commander, Training and Education Command, personal interview by author, February 7, 2005, Fort Monroe, VA; Dixon et al., *Companycommand: Unleashing the Power of the Army Professions* (West Point, NY: Center for the Advancement of Leader Development and Organizational Learning, 2005); General Gordon Sullivan, USA, personal interview by author, February 25, 2005, Arlington, VA.

to be reviewed and discussed. Its use was *permitted* to be studied but rarely *promoted* in education. Instead, formal (albeit indirect) exposure to such new doctrine was mostly accomplished through its incorporation into training scenarios at the CTCs (scenario developers and observer controllers at the JRTC have well-worn volumes of these field manuals at their desks) and by the active "pulling" of this information from the websites and CALL by individual commanders as they attempted to prepare their units for deployment and then cope in the field. This is an example of how an institution becomes aware of formal doctrine even though it may be rarely read or discussed in the classroom.

Starting in 2001, conflicts in Iraq and Afghanistan forced dramatic changes to incorporate the problems of stability operations throughout the Army's PME curriculum.[7] Likewise, the Marine Corps, which had consistently done a slightly better job covering MOOTW in its schools, began to initiate a new cultural education program to better prepare career marines for complex missions abroad.[8] These changes, which were suggested as a result of various studies of recent operations, including interviews with veterans returning from Iraq, suggest a positive shift in the learning culture of the services. Such a shift in culture reflects how well-designed internal structures and processes can act as a catalyst for bottom-up cultural change. This is especially critical during wartime when new experiential knowledge is being absorbed.

The learning system described here, which is deliberately based on learning theory, has enhanced the military's capacity for institutional learning. More study and observation in the coming years are required to determine if the late educational changes will be sustained and if they reflect true shifts in organizational culture. Still, the changes described in this research clearly indicate generational learning as a result of operational experience—at least at the tactical level.

One potential critique of this thesis is that this sort of learning does

---

7. Brigadier General Volney J. Warner, USA, Deputy Commandant, Command and General Staff College, "Evolving and Adapting to Meet Challenges and Exploit Opportunities: A Presentation for Ms. Tammy Schultz" (Fort Leavenworth, KS, April 2005).

8. Barak Salmoni, Director, Marine Corps Center for Advanced Operational Cultural Learning, presentation to Marine Corps conference, "Irregular Warfare: Creating Stability in an Unstable World," Center for Emerging Threats and Opportunities, Quantico, VA, July 12, 2005; Colonel John Toolan, USMC, Director, Marine Corps Command and Staff College, presentation to Marine Corps conference, "Irregular Warfare: Creating Stability in an Unstable World," Center for Emerging Threats and Opportunities, Quantico, VA, July 12, 2005; Warner, "Evolving and Adapting to Meet Challenges and Exploit Opportunities."

not reflect widespread substantive change.[9] It is true that peace operations training at the unit level throughout the 1990s only reflected a small percentage of the overall training experience of most troops. Individual training such as basic training at the branch schools as well as much of the units' other formal training still emphasized traditional operations. Still, the real-world experiential learning and the training that was derived from it were significant. Over 230,000 troops participated in such missions and conducted predeparture training from 1992 to 2003.[10] Most important for institutional learning and the capacity for the military to adapt, the changes made at the CTCs provided institutional seeds that have been allowed to grow.

As the rest of the military (including the Marine Corps) looked to rapidly adapt its training for the massive stabilization and reconstruction operations in Iraq, the CTC curricula, scenarios, and lessons learned throughout the 1990s provided the basis for their programs. With respect to theories of military change, these seeds demonstrate internal, rather than external, catalysts for change. The gradual pollination of these seeds is evident if one compares substantive changes from 1989 to the present—in officer development as well as the tactics and learning processes in place for Iraq today.

Post–Cold War Generational Learning: Officer Career Comparison

The substantive post–Cold War shifts in the overall learning system are best reflected by a comparison of three generations of career officers. Consider the officer commissioning classes of 1975, 1985, and 1995. The class of 1975 would become the brigade commanders and senior staff officers for the First Gulf War in 1990. These officers would have spent most of their careers in training, rather than in real-world operations. They would have been the first post-Vietnam generation to benefit from the changes put in place by General Gorman and his contemporaries. By 1990, regular branch officers would have attended a rotation or two at the new National Training Center, where they would have practiced the new AirLand Battle Doctrine. They would have had nearly zero

---

9. Thank you to Lieutenant Colonel Richard Lacquement, Jr., PhD, for presenting this critique. See also Richard Lacquement, *Shaping American Military Capabilities after the Cold War* (London: Praeger, 2003).

10. Colonel (Ret.) George Oliver, U.S. Army, former Director of the Army Peacekeeping Institute, "U.S. Peace Operations since 1981: Troop Strengths" (Washington, DC: Center for Military History, 2005).

instruction on counterinsurgency or MOOTW, either in their professional military education or in their training. Still, by the time they became field-grade officers and as they approached the rank of colonel, some of them might have led battalions and brigades in Grenada, Panama, and Somalia—experiences that would easily have raised questions about the adequacy of their doctrine, education, and training. The class of 1975 would become the senior military leaders of the 1990s and into the 21st century. One of the most prominent of this generation is David Petreaus, who, as a lieutenant general in 2006 and commander of the Army's Combined Arms Center, would spearhead the publication of a new counterinsurgency (COIN) manual based on a combination of recent operational experience and classic COIN theory. To ensure wide dissemination and organizational change, he would institute significant revisions to the educational system to account for this new doctrine. Later, as a four-star general and commander of the Multi-National Force–Iraq (MNF-I) in 2007, General Petreaus would attempt to apply this doctrine in the field.

The next generation of officers, the class of 1985, would experience a dramatic mismatch between what they were learning and what they were doing operationally. As cadets in the early 1980s, their West Point or ROTC education would still not have addressed counterinsurgency or MOOTW. Operationally, some may have participated as junior officers in the First Gulf War, but the majority of their careers would be spent conducting MOOTW from Somalia and Haiti to the Balkans, Afghanistan, and Iraq. This generation of officers would be the battalion commanders and the brigade and division staff officers for the 2003 invasion of Iraq and its multiyear aftermath.

Throughout the 1990s and especially after 2003, the gap between operational reality and training for the class of 1985 would become slightly smaller. This was especially true for the Light Infantry, the Special Operations Forces, and others whose experience at the JRTC (opened in 1992) was already focused on the lower-intensity and counterinsurgency missions. For officers of all branches deploying to the Balkans, predeparture mission rehearsal exercises (MREs) at both the JRTC and the CMTC would have reflected the latest problems being encountered in Bosnia and Kosovo. Predeployment training before the Iraq invasion focused on traditional major combat operations, but by the time the Iraq insurgency began in the fall of 2003, the scenarios honed at the JRTC and the CMTC had been adopted by the larger National Training Center in California, so that all units rotating through these centers were being trained on the latest tactics being fed from Iraq and were using many techniques learned throughout the 1990s. As these officers and their immediate subordinates rotated back to the United States and entered mid- and senior-level PME, they would be highly critical of the out-of-touch curricula and would push for change in the education system. This change would affect the next generation, for whom real-world operations, education, and training would have begun to align.

The third generation of officers, the class of 1995, would have cut their teeth on operations in the Balkans. This would be the generation to start online communities of practice, such as CompanyCommand.com, which they would insist the formal institution support. Moreover, in contrast to their commanders of the previous generation, their education at West Point would have begun to address counterinsurgency and MOOTW. For infantry units, these topics would have been reinforced in their basic branch course and then again at their JRTC rotations. Due to the subtle post–Cold War changes based on operational experience and the lessons-learned processes, these topics have become recurring themes for this group throughout their careers. They studied them at West Point, practiced them at their Basic Infantry course and at the JRTC, and learned about peace operations either firsthand in Kosovo or secondhand from their commanders and NCOs who had served in Bosnia. As one officer of this generation put it, "peacekeeping, counterinsurgency, and low-intensity conflict were the steady state [for us], and 'big war' was the spice."[11]

In sum, there is no doubt that commanders serving today have benefited at the tactical level from the experiential learning of the 1990s and from the new and improved post-Vietnam learning system. The military that entered Baghdad in 2003 was neither the military of the Vietnam era nor the same military that General Petraeus would command during the Surge in 2007. The military had adapted greatly to the exigencies of the war, yet political success would not necessarily follow quickly.

The case of Iraq demonstrates that military learning, no matter how comprehensive or how swift, is necessary but not always sufficient for strategic success in complex stabilization and reconstruction operations or counterinsurgency. When the strategic objectives require success in the "fourth block," a realm that often involves a number of nonmilitary actors and an environment in which "lessons" are slow to emerge, the military's system of rapid tactical learning has limited utility.

---

11. Captain Scott Sonsalla, USA, Address given at Army War College conference, Carlisle, PA, March 13, 2004.

Appendix: Key Terms and Conceptual Confusion

American military doctrine and theory has struggled with terminology for decades. Below are explanations of the most common terms in use today[1] and brief discussions of the controversies surrounding them.

Counterinsurgency (COIN)

- DoD: "Those military, paramilitary, political, economic, psychological, and civic actions taken by a government to defeat an insurgency."[2]
- State Department: "Comprehensive civilian and military efforts taken to simultaneously defeat and contain insurgency and address its root causes."[3]
- These two "official" definitions reflect the fact that countering an insurgency requires a strategy tailored to the particular nature of the insurgency. As discussed in this book, this entails a comprehensive assessment of the root causes as well as the tactics, techniques, and strategy of the insurgents.

Counterterrorism (CT)

- DoD: "Operations that include the offensive measures taken to prevent, deter, preempt, and respond to terrorism."

---

1. All Department of Defense definitions are available online at http://www.dtic.mil/doctrine/jel/doddict/.
2. DoD Dictionary version, October 17, 2008, http://www.dtic.mil/doctrine/jel/doddict/. All citations of DoD in the appendix came from this source unless specified otherwise.
3. *U.S. Government Counterinsurgency Guide* (State Department Bureau of Political-Military Affairs, January 2009), www.state.gov/t/pm/ppa/pmppt.

- CT is different from COIN in that terrorists are not necessarily in a competition for control of the population against the local or regional governing authority. CT operations are thus offensive operations, focused less on a competition for governance and more on undermining and disabling the terrorist *network*. In some Commonwealth countries, there is a distinction made between "antiterrorism," which is defensive in nature, and "counterterrorism," which is more offensive, disruptive, or preventive, as described here.

Foreign Internal Defense (FID)

- DoD: "Participation by civilian and military agencies of a government in any of the action programs taken by another government or other designated organization to free and protect its society from subversion, lawlessness, and insurgency."
- See also "IDAD" below. IDAD emphasizes the preventative steps a state takes to protect itself from such threats.

Guerrilla Warfare

- DoD: "Military and paramilitary operations conducted in enemy-held or hostile territory by irregular, predominantly indigenous forces."
- *Guerrilla* is derived from the Spanish word for "war," *Guerra,* with the suffix for "little," *illa.* See "Small Wars" below.

Insurgency

- DoD: "An organized movement aimed at the overthrow of a constituted government through use of subversion and armed conflict."
- State Department: "The organized use of subversion and violence to seize, nullify, or challenge political control of a region."

Internal Defense and Development (IDAD)

- DoD: "The full range of measures taken by a nation to promote its growth and to protect itself from subversion, lawlessness, and insurgency. It focuses on building viable institutions (political, economic, social, and military) that respond to the needs of society."
- *IDAD* is a term used since the Cold War and, in many respects, can be seen as the more preventative side of COIN or FID. COIN is conducted

in response to, rather than in anticipation of, an insurgent threat (though early intervention is preferred). FID emphasizes the role of an intervening power in supporting the security elements of a state.

Irregular Warfare (IW)

- DoD: "A violent struggle among state and non-state actors for legitimacy and influence over the relevant population(s). Irregular warfare favors indirect and asymmetric approaches, though it may employ the full range of military and other capacities, in order to erode an adversary's power, influence, and will."
- This term is increasingly popular in the U.S. military, especially the Marine Corps and the Special Operations community. Others, including many in the U.S. Army, object to the term because of its conceptual confusion and because nonmilitary partners object to having their mission sets recrafted under the "war" terminology. This is especially the case among diplomats who understand that with respect to "diplomacy" and international law, *war* has very specific meanings. Other objections in the military community focus on the intellectual difficulty in categorizing too many things under one "umbrella" term (such as *MOOTW*) and in defining something by what it is not (i.e., not "regular").

Kinetic

- *Kinetic* is a nondoctrinal term used increasingly today to differentiate the more violent, direct, enemy-focused military operations from the more subtle, population-focused, often civilian-led operations conducted in counterinsurgency and stability operations. A key differentiating element of "kinetic" actions is that they can be potentially lethal.[4]

Low-Intensity Conflict (LIC)

- No longer listed in the DoD Dictionary
- Previously, the U.S. Army defined LIC as "a political-military confrontation between contending states or groups below conventional war and above the routine, peaceful competition among states. It frequently involves protracted struggles of competing principles and ideologies. Low-intensity conflict ranges from subversion to the use of the armed forces.

---

4. Thank you to David Kilcullen for help in clarifying the common use of this term.

It is waged by a combination of means, employing political, economic, informational, and military instruments. Low-intensity conflicts are often localized, generally in the Third World, but contain regional and global security implications."[5]

- In the office of the secretary of defense in the Pentagon, there is still an assistant secretary of defense for special operations and low-intensity conflict and integrating capabilities (ASD for SOLIC/IC). This office was congressionally mandated in the 1980s, and it would take a revision by Congress to change it. Currently, the ASD for SOLIC/IC oversees the development of capabilities for stability operations, counterinsurgency, and irregular warfare.

Military Operations other than War (MOOTW)

- No longer listed in the DoD Dictionary
- *MOOTW* was a doctrinal term used in the 1990s by the American military to describe a large number of what they considered nontraditional tasks. The 1995 version of Joint Publication 3-07 *Military Operations other than War* defined the term as "operations that encompass the use of military capabilities across the range of military operations short of war. These military actions can be applied to complement any combination of the other instruments of national power and occur before, during, and after war."[6] The manual goes on to explain that these missions can be either combat or noncombat in nature. They include categories as diverse and dangerous as "arms control; combating terrorism; . . . support to counterdrug operations; enforcement of sanctions/maritime intercept operations; enforcing exclusion zones; ensuring freedom of navigation and overflight; humanitarian assistance; military support to civil authorities; nation assistance/support to counterinsurgency; noncombatant evacuation operations; peace operations; protection of shipping; recovery operations; show of force operations; strikes and raids; and support to insurgency."[7] The diverse range of operations included under the category of MOOTW, combined with the military's tradition of focusing primarily on major theater warfare, led to a popular joke among officers that *MOOTW* really means "military operations other than *what we want to do.*"

---

5. *U.S. Army Field Manual 100-20. Military Operations in Low Intensity Conflict* (1990). Replaced by *FM 3-24 Counterinsurgency* (2006) and *FM 3-07 Stability Operations* (2008).

6. Department of Defense, *Joint Publication 3-07: Joint Doctrine for Operations Other than War* (Washington, DC: U.S. Government Printing Office, June 16, 1995).

7. Ibid.

Small Wars

- No longer listed in the DoD Dictionary
- *Small wars* was a term used to describe British colonial clashes in Asia and Africa in the late 19th and early 20th centuries. It was made popular at the time by the publication of *Small Wars: Their Principles and Practice* in 1896 by C. E. Callwell, a British colonel (later a major general) and veteran of the Second Afghan and Boer wars.[8] The term was used by the American Army to describe its own clashes with Native Americans in the 1800s. The Marine Corps used the term to describe their experience in the Caribbean "Banana Wars" of the 1930s and later published the *Small Wars Manual* in 1940.[9] Today, the popular Small Wars Journal website, which is run by retired U.S. marines, publishes articles and discussions on counterinsurgency, stability operations, and other matters relating to 21st-century conflict, as well as hosting discussions by the key thought leaders in the field.

Stability Operations

- DoD: "An overarching term encompassing various military missions, tasks, and activities conducted outside the United States in coordination with other instruments of national power to maintain or reestablish a safe and secure environment, provide essential governmental services, emergency infrastructure reconstruction, and humanitarian relief."
- Pentagon Policy: In 2005, a new DoD directive issued by the deputy secretary of defense placed greater emphasis on the military's role in stability operations. According to DoDD 3000.05 *Military Support for Stability, Security, Transition, and Reconstruction (SSTR) Operations,* "Stability operations are conducted to help establish order that advances U.S. interests and values. The immediate goal often is to provide the local populace with security, restore essential services, and meet humanitarian needs. The long-term goal is to help develop indigenous capacity for securing essential services, a viable market economy, rule of law, democratic institutions, and a robust civil society."

---

8. C. E. Callwell, *Small Wars: Their Principles and Practice,* 3rd ed. (Lincoln: University of Nebraska Press, 1996).

9. USMC, *Small Wars Manual: United States Marine Corps* (1940; repr., Sunflower University Press, 1996).

Unconventional Warfare (UW)

- DoD: "A broad spectrum of military and paramilitary operations, normally of long duration, predominantly conducted through, with, or by indigenous or surrogate forces who are organized, trained, equipped, supported, and directed in varying degrees by an external source. It includes but is not limited to guerrilla warfare, subversion, sabotage, intelligence activities, and unconventional assisted recovery."
- UW has come to be seen as more offensive in nature and as a type of warfare that usually involves nonstate actors. UW is often conflated with GW or IW, highlighting the continued conceptual confusion and debate among theorists, policymakers, and practitioners.

Bibliography

Achenback, Joel. "George Washington's Western Frontier." *Washington Post Magazine*, June 6, 2004, 11–15.
Ackerman, Spencer. "The Colonels and 'the Matrix.'" *Washington Independent*, July 28, 2008. First article in the series The Rise of the Counterinsurgents. http://washingtonindependent.com/view/the-colonels-and-the.
Adelman, Ken. "Cakewalk in Iraq." *Washington Post*, February 13, 2002, A27.
Agoglia, Colonel John, U.S. Army. Personal interview by author. July 15, 2007. Arlington, VA.
Air Land Sea Application Center. *Field Manual 3-07.31/MCWP 3-33.8/AFTTP(I) 3-2.40: Multi-Service Tactics, Techniques, and Procedures for Conducting Peace Operations*. Langley Air Force Base, VA, October 2003.
Allard, C. Kenneth. *Somalia Operations: Lessons Learned*. Washington, DC: National Defense University Press, 1995.
Allison, Graham. *Essence of Decision: Explaining the Cuban Missile Crisis*. Boston: Little, Brown, 1971.
Allison, Graham. *Essence of Decision: Explaining the Cuban Missile Crisis*. 2nd ed. New York: Longman, 1999.
Altfeld, Michael, and Gary Miller. "Sources of Bureaucratic Influence: Expertise and Agenda Control." *Journal of Conflict Resolution* 28, no. 4 (1984).
Angevine, Robert, Christine Grafton, and Jeff McKitrick. *Changing Military Cultures II: Report Prepared for the Office of Net Assessment, Office of the Secretary of Defense*. Strategic Assessment Center, SAIC, October 17, 2003.
Argyris, Chris, and Donald Schon. *Organizational Learning II: Theory, Method, and Practice*. Reading, MA: Addison-Wesley, 1996.
Army, Department of the. *Annual Historical Summaries*. Washington, DC: Center for Military History, 1983–99.
Army, Department of the. *Field Manual 3-0 Operations (Formerly FM 100-5)*. Washington, DC, 2001.
Army, Department of the. *Field Manual 3-06 Urban Operations (Formerly FM 90-10)*. Washington, DC, June 1, 2003.
Army, Department of the. *Field Manual 100-5 Operations*. Washington, DC, 1982.
Army, Department of the. *Field Manual 100-5 Operations*. Washington, DC, 1986.
Army, Department of the. *Field Manual 100-5 Operations*. Washington, DC, 1993.

Army, Department of the. *Field Manual 100-23 Peace Operations*. Washington, DC, 1994.
Army, Department of the, and Department of the Air Force. *FM 100-20/AFP 3-20 Low Intensity Conflict*. Washington, DC, 1990.
Army, U.S. *FM 3-07 (FM 100-20) Stability and Support Operations*. Washington, DC: Headquarters, U.S. Army, February 20, 2003.
Army, U.S. *Small Wars and Punitive Expeditions*. Vol. 2, *Tactics*. Fort Benning, GA: U.S. Army Infantry School, 1925–26.
Army news release. "Army Announces Fy05 and Fy06 Modular Brigade Force Structure Decisions." *Army Public Affairs,* July 23, 2004. http://www.globalsecurity.org/military/library/news/2004/07/mil-040723-army01.htm.
Arnold, Steven L. "Somalia—an Operation Other than War." *Military Review* (December 1993).
Aspin, Les, Secretary of Defense. "Bottom-up Review." Washington, DC, October 1993.
Atkinson, Rick. *An Army at Dawn: The War in Africa, 1942–1943*. Liberation Trilogy, vol. 1. New York: Holt, 2007.
Avant, Deborah D. "The Institutional Sources of Military Doctrine: Hegemons in Peripheral Wars." *International Studies Quarterly* 37 (1993): 409–30.
Avant, Deborah. *Political Institutions and Military Change: Lessons from Peripheral Wars*. Ithaca: Cornell University Press, 1994.
Avant, Deborah, and James Lebovic. "U.S. Military Attitudes towards Post-Cold War Missions." *Armed Forces and Society* 27, no. 1 (2000): 37–56.
Aylwin-Foster, Brigadier Nigel, British Army. "Changing the Army for Counterinsurgency Operations." *Military Review* (November–December 2005).
Bailey, Colonel Ronald, USMC, Commander, 2nd Regiment, 2nd Marine Division. Personal interview by author. September 23, 2003. Arlington, VA.
Bain, David Haward. *Sitting in Darkness: Americans in the Philippines*. Boston: Penguin, 1986.
Baird, Lloyd, John C. Henderson, and Stephanie Watts. "Learning from Action: An Analysis of the Center for Army Lessons Learned (CALL)." *Human Resource Management* 36, no. 4 (Winter 1997): 385–95.
Baker, Bonnie. "The Origins of the *Posse Comitatus*." *Air and Space Power Chronicles* (November 1999). http://www.airpower.maxwell.af.mil/airchronicles/cc/baker1.html.
Baker, James A., and Lee H. Hamilton. "The Iraq Study Group Report." Washington, DC: United States Institute of Peace, 2006. http://www.usip.org/isg/iraq_study_group_report/report/1206/index.html.
Baker, Peter, and Robin Wright. "To 'War Czar,' Solution to Iraq Conflict Won't Be Purely Military." *Washington Post,* May 17, 2007, A13.
Barnett, Thomas P. M. "Donald Rumsfeld: Old Man in a Hurry." *Esquire,* July 2005.
BDM Corporation. *A Study of the Strategic Lessons Learned in Vietnam: Omnibus Executive Summary*. Washington, DC: BDM Corporation (sponsored by the U.S. Army War College), 1980.
Benson, Colonel Kevin, USA, Director, School of Advanced Military Planning. Personal interview by author. March 9, 2004. Fort Leavenworth, KS.

Bickel, Keith B. *Mars Learning: The Marine Corps Development of Small Wars Doctrine, 1915–1940.* Boulder: Westview Press, 2001.
Bickel, Keith. Personal interview by author. September 27, 2004. McLean, VA.
Binnendijk, Hans, and Stuart E. Johnson. *Transforming for Stabilization and Reconstruction.* Washington, DC: National Defense University Press, 2004.
Birtle, Andrew J. *U.S. Army Counterinsurgency and Contingency Operations Doctrine 1865–1941.* Washington, DC: Center of Military History, 1998.
Bois, W. E. Burghardt Du. "The Freedmen's Bureau." *Atlantic Monthly* (March 1901).
Bolger, Daniel P. *The Battle for Hunger Hill: The 1st Battalion, 327th Infantry Regiment at the Joint Readiness Training Center.* Novato, CA: Presidio, 1997.
Bonn, Lieutenant Colonel (Ret.) Keith E., USA, and Master Sergeant (Ret.) Anthony E. Baker, USAR. *Guide to Military Operations Other than War: Tactics, Techniques, and Procedures for Stability and Support Operations.* Mechanicsburg, PA: Stackpole Books, 2000.
Boot, Max. "Reconstructing Iraq: With the Marines in the South and the 101st Airborne in the North." *Weekly Standard* (September 15, 2003).
Boot, Max. *The Savage Wars of Peace: Small Wars and the Rise of American Power.* New York: Basic Books, 2002.
Bowden, Mark. *Black Hawk Down: A Story of Modern War.* New York: Atlantic Monthly Press, 1999.
Bremer, Paul. "Coalition Provisional Authority Order Number 1: De-Ba'Athification of Iraqi Society." Baghdad, May 12, 2003.
Bremer, Paul. "Coalition Provisional Authority Order Number 2: Dissolution of Entities." Baghdad, May 23, 2003.
Brinsfield, John. "The Military Ethics of General William T. Sherman." *Parameters* 12 (June 1982): 36–48.
Brown, Lieutenant Colonel Kevin, USA, scenario developer, Joint Readiness Training Center. Personal interview by author. March 27, 2004. Fort Polk, LA.
Brune, Lester H. *The United States and Post Cold War Interventions.* Claremont, CA: Regina Books, 1998.
Builder, Carl H. *The Masks of War: American Military Styles in Strategy and Analysis.* Rand Corporation Research Study. London: Johns Hopkins University Press, 1989.
Bullard, Robert L. "Military Pacification." *Journal of the Military Service of the United States* 46, no. 143 (1910): 1–24.
Burke, Colonel (Ret.) Michael D., USA, "FM 3-0: Doctrine for a Transforming Force." *Military Review* (March–April 2002): 91–97.
Burke, Michael, member, writing team *FM 100-5*, 1996–2001, *FM 3-0* 2001 and 2005 editions. Personal interview by author. February 3, 2005. Via telephone to Combined Arms Doctrine Division, Fort Leavenworth, KS.
Butler, Brigadier General Smedley D. Introduction to *War Is a Racket: The Antiwar Classic by America's Most Decorated Soldier,* edited by Adam Parfrey. 1935. Reprint, Los Angeles: Feral House, 2003.
Byrnes, General Kevin, USA, Commander, Training and Education Command. Personal interview by author. February 7, 2005. Fort Monroe, VA.
Callwell, Colonel C. E. *Small Wars.* 3rd ed. London: E. P. Publishing, 1976.

Capps, Steve, editor *FM 3-0* 2001 and 2005 editions. Personal interview by author. February 8, 2005. Via telephone and follow-up e-mails to Combined Arms Doctrine Division, Fort Leavenworth, KS.

Cassidy, Robert M. *Peacekeeping in the Abyss: British and American Peacekeeping Doctrine and Practice after the Cold War.* London: Praeger, 2004.

Cassidy, Colonel Robert. "Prophets or Praetorians? The Uptonian Paradox and the Powell Corollary." *Parameters* (Autumn 2003): 130–43.

Cassidy, Robert M. "Why Great Powers Fight Small Wars Badly." *Military Review* (2000).

Center for Army Lessons Learned (CALL). *CALL Handbook: U.S. Army and U.S. Marine Corps Tactics, Techniques, and Procedures for Conducting Peace Operations.* Fort Leavenworth, KS: Center for Army Lessons Learned, 1999.

Center for Army Lessons Learned (CALL). "Foreword." *Combat Training Center Quarterly Bulletin,* fourth quarter. Fort Leavenworth, KS: Center for Army Lessons Learned, 2002.

Center for Army Lessons Learned (CALL). *Operations Other than War.* Vol. 4, *Peace Operations.* Fort Leavenworth, KS: U.S. Army Command and General Staff College, 1993.

Chapman, Anne W. *The Army's Training Revolution: 1973–1990, an Overview.* Fort Monroe, VA, and Washington, DC: Office of the Command Historian, United States Training and Doctrine Command and the Center of Military History, 1994.

Cheney, Richard, Secretary of Defense. "The Gulf War: A First Assessment." Presented at the Soref Symposium, Washington Institute for Near East Policy, Washington, DC, 1991.

Chiarelli, Major General Peter, USA, Commander, First Cavalry Division. Presentation on Operations in Baghdad, 1st Cavalry. Powerpoint briefing, 2005.

Chiarelli, Major General Peter, USA, Commander, First Cavalry Division, and Major Patrick R. Michaelis, USA. "Winning the Peace: The Requirement for Full Spectrum Operations." *Military Review* (July–August 2005).

Chura, Michael, editor of *FM 3-07*. E-mail correspondence with author. February 14, 2005.

Chura, Mike, doctrine writer, Combined Arms Doctrine Division. Personal interviews by author. February 2005. Via telephone and e-mail. Orlando, FL.

Clancy, Tom, General (Ret.) Anthony Zinni, USMC, and Tony Koltz. *Battle Ready.* New York: Putnam, 2004.

Clausewitz, Carl von. *On War.* Edited by Michael Howard and Peter Paret. Princeton: Princeton University Press, 1984.

Clinton, President William Jefferson. "Military Intervention in Serbia 1999, Statement by the President to the Nation." Oval Office, Washington, DC, March 24, 1999.

Clinton, President William Jefferson. "Presidential Decision Directive-25: Clinton Administration Policy on Reforming Multilateral Peace Operation." Washington, DC, 1994.

Clinton, William Jefferson. *A Strategy of Engagement and Enlargement.* Washington, DC: White House, February 1996.

"Clinton Administration Policy on Reforming Multilateral Peace Operations

(PDD 25), Executive Summary." Bureau of International Organizational Affairs, U.S. Department of State, February 22, 1996.

Cohen, Eliot. "Constraints on America's Conduct of Small Wars." In *Conventional Forces and American Defense Policy*, edited by Steven E. Miller. Princeton: Princeton University Press, 1986.

Cohen, Eliot. "A Hawk Questions Himself as His Son Goes to War." *Washington Post*, July 10, 2005, B1.

Cole, Ronald H. *Operation Just Cause: The Planning and Execution of Joint Operations in Panama*. Joint History Office, Office of the Chairman of the Joint Chiefs of Staff, 1995.

Coles, Harry L., and Albert K. Weinberg. *Civil Affairs: Soldiers Become Governors, U.S. Army in World War Two*. Washington, DC: U.S. Army Center for Military History, 1992.

Collins, Joseph, former Deputy Assistant Secretary for Stability Operations. Numerous personal interviews by author. 2004–5. Washington, DC.

Combat Studies Institute, U.S. Army Command and Staff College. "CSI Report No. 14: Sixty Years of Reorganizing for Combat: A Historical Trend Analysis." Fort Leavenworth, KS, January 2000.

Commandant, U.S. Army Infantry School. *White Paper: The Application of Peace Enforcement (PE) at Brigade and Battalion, GPO*. Fort Benning, GA, August 31, 1993.

"Commission on Roles and Missions of the Armed Forces, Report to Congress, the Secretary of Defense, and the Chairman of the Joint Chiefs of Staff." Washington, DC, May 24, 1995.

Congressional Budget Office. "Making Peace while Staying Ready for War: The Challenge of U.S. Military Participation in Peace Operations." Washington, DC, 1999.

Corbin, Marcus. "Revive Combined Action Platoons for Iraq." *Defense News*, Oct. 11, 2004. http://www.defensenews.com/story.php?F=452519&C=commentary.

Crane, Conrad. "Avoiding Vietnam: The U.S. Army's Response to Defeat in Southeast Asia." *Strategic Studies Institute* (2002).

Crane, Conrad. "Landpower and Crises: Army Roles and Missions in Smaller-Scale Contingencies during the 1990's." *Strategic Studies Institute* (2001).

Crawford, Oliver. *The Door Marked Malaya*. London: Rupert Hart-Davis, 1958. Reprinted in 1989 as *FMFRP 12-28 USMC*.

Cyert, Richard M., and James G. March. *A Behavioral Theory of the Firm*. 2nd ed. Malden, MA: Blackwell, 1992.

Daft, R. L., and K. E. Weick. "Toward a Model of Organizations as Interpretation Systems." *Academy of Management Review* 9, no. 2 (1984): 284–95.

Davidson, Janine. "A Citizen Check on War." *Washington Post*, Op-ed, November 16, 2003.

Davidson, Janine. "Doing More with Less? The Politics of Readiness and the U.S. Use of Force." M.A. thesis, University of South Carolina, 2002.

Davidson, Janine. "Nation Building Nexus: Post Conflict Reconstruction and Military Doctrine for Stability Operations." Paper presented at the International Studies Association, Southern Region, Annual Conference, Columbia, SC, October 23, 2004.

Deady, Timothy K. "Lessons from a Successful Counterinsurgency: The Philippines, 1899–1902." *Parameters* 35, no. 1 (2005).
Decker, Marvin. Personal interview by author. March 8, 2004. Center for Army Lessons Learned, Fort Leavenworth, KS.
Defense, Department of. "Active Duty Military Personnel Strengths by Regional Area and by Country (309a)." Washington, DC: Washington Headquarters Services, Directorate for Information Operations and Reports, September 30, 1997.
Defense, Department of. *Joint Publication 3-07: Joint Doctrine for Operations Other than War.* Washington, DC: U.S. Government Printing Office, June 16, 1995.
Defense, Department of. *Joint Publication 3-07.3: Joint Tactics, Techniques, and Procedures for Peace Operations.* Washington, DC: U.S. Government Printing Office, February 12, 1999.
Defense, Department of. *Joint Publication 3-08: Joint Doctrine for Interagency Coordination.* Vol. 1. Washington, DC: U.S. Government Printing Office, October 9, 1996.
Defense, Department of. *Joint Publication 3-08: Joint Doctrine for Interagency Coordination.* Vol. 1 (draft). Washington, DC: U.S. Government Printing Office, April 26, 2004.
Deutschmann, Rod. "Top Military, Civilian Officials Discuss Future of Humanitarian Operations." *Navy Wire Service,* Navy Public Affairs Library, April 27, 1995.
Dilegge, Dave. Personal correspondence with the author, Washington, DC, 2008.
Dixon, Nancy M., Nate Allen, Tony Burgess, Pete Kilner, and Steve Schweitzer. *Companycommand: Unleashing the Power of the Army Professions.* West Point, NY: Center for the Advancement of Leader Development and Organizational Learning, 2005.
Downie, Richard Duncan. *Learning from Conflict: The U.S. Military in Vietnam, El Salvador, and the Drug War.* Westport, CT: Praeger, 1998.
Downs, Anthony. *Inside Bureaucracy.* Boston: Little, Brown, 1967.
Dumas, Captain Wess, USA. Personal interview by author. October 13, 2004. Fort Campbell, KY.
Dunlop, Colonel Charles J., USAF. "Welcome to the Junta: The Erosion of Civilian Control of the U.S. Military." *Wake Forest Law Review* 341 (1994).
Echevarria, Antulio J. "Toward an American Way of War." Strategic Studies Institute Monograph. Carlisle, PA: U.S. Army War College, 2004. http://www.carlisle.army.mil/ssi/.
Elman, Colin, and Miriam Fendius Elman. "Diplomatic History and International Relations Theory: Respecting Difference and Crossing Boundaries." *International Security* 22, no. 1 (1997): 5–21.
Fallows, James. "Blind into Baghdad." *Atlantic* (January–February 2004).
Fastabend, Lieutenant Colonel David, USA. *A General Theory of Conflict: Bosnia, Strategy, and the Future.* Hoover Institution, Strategy Research Project, Stanford University, 1996.
Fastabend, Brigadier General David, Deputy Director, Army TRADOC Futures Center. Personal interview by author. February 8, 2005. Fort Monroe, VA.
Feaver, Peter. *Armed Servants.* Cambridge, MA: Harvard University Press, 2003.

Feaver, Peter, and Richard H. Kohn. "The Gap: Soldiers, Civilians, and Their Mutual Misunderstanding." *National Interest* 61 (Fall 2000): 29.

Feaver, Peter D., and Richard H. Kohn, eds. *Soldiers and Civilians: The Civil-Military Gap and American National Security.* Cambridge, MA: MIT Press, 2001.

Felix, Lieutenant Colonel Kevin, USA, Commander, 2-320 Field Artillery Battalion. Personal interview by author. October 13, 2004. Fort Campbell, KY.

Fitzgerald, Colonel B. R. *Commander's Intent, Task Force Falcon, 3A.* May 2001.

Flournoy, Michele. *Interagency Strategy and Planning for Post-Conflict Reconstruction.* Draft white paper produced for the Post-Conflict Reconstruction Project. Washington, DC: Center for Strategic and International Studies, March 27, 2002.

Flournoy, Michelle, former Deputy Assistant Secretary of Defense for Strategy. Personal interview by author. 2003. Center for Strategic and International Studies, Washington, DC.

Ford, Allen S. "The Small Wars Manual and Marine Corps Military Operations Other than War Doctrine." In *Fort Leavenworth Papers.* Ft. Leavenworth, KS, 2003.

Forman, Johanna Mendelson, and Michael Pan. "Filling the Gap: Civilian Rapid Response Capacity for Post-Conflict Reconstruction." In *Winning the Peace,* edited by Robert Orr. Significant Issues Series. Washington, DC: Center for Strategic and International Studies, 2004.

Foster. "Pentagon Peacekeeping 101." *Milwaukee Journal Sentinel,* July 14, 2003, A08.

Gates, John. "The Alleged Isolation of U.S. Army Officers in the Late-19th Century." *Parameters: Journal of the U. S. Army War College* 10 (1980): 32–45.

Gates, John. "Indians and Insurrectos: The U.S. Army's Experience with Insurgency." *Parameters* 13 (1983): 56–68.

Gates, John. *Schoolbooks and Krags: The United States Army in the Philippines, 1898–1902.* Westport, CT: Greenwood Press, 1980.

Gavrilis, James A. "The Mayor of Ar Rutbah." *Foreign Policy* (November–December 2005).

General Accounting Office. "Army Training: National Training Center's Potential Has Not Been Realized (GAO/NSIAD 86-130)." Washington, DC, July 23, 1986.

General Accounting Office. "Military Operations: Impact of Operations Other than War on the Services Varies (GAO/NSIAD 99-69)." Washington, DC, May 1999.

General Accounting Office. "Military Training: Potential to Use Lessons Learned to Avoid Past Mistakes Is Largely Untapped (GAO/NSIAD 95-152)." Washington, DC, 1995.

General Accounting Office. "Peace Operations: Effect of Training, Equipment, and Other Factors on Unit Capability (GAO/NSIAD-96-14)." Washington, DC, October 1996.

"General: Training Key to Worldwide Missions." *Times-Picayune,* August 15, 1994, B8.

Gentile, Gian. "Eating Soup with a Spoon: Missing from the New Coin Manual's Pages Is the Imperative to Fight." *Armed Forces Journal* (September 2007).

George, Alexander. *Bridging the Gap: Theory and Practice in Foreign Policy.* Washington, DC: U.S. Institute of Peace, 1993.
George, Alexander. "Case Studies and Theory Development: The Method of Structured, Focused Comparison." In *Diplomacy: New Approaches in History, Theory, and Policy,* edited by Paul Gordon Lauren. New York: Free Press, 1979.
Gladwell, Malcolm. *Blink: The Power of Thinking without Thinking.* New York: Little, Brown, 2005.
Goldwater-Nichols Department of Defense Reorganization Act of 1986. 99-433. October 1, 1986.
Goodman, Sherri W., Deputy Undersecretary of Defense for Environmental Security. Presentation at the Women in International Security Summer Symposium, Pentagon, Arlington, VA, June 20, 2000.
Gordon, Michael, and Bernard Trainer. *Cobra II: The Inside Story of the Invasion and Occupation of Iraq.* New York: Pantheon, 2006.
Graham, Bradley. "New Twist for U.S. Troops: Peace Maneuvers." *Washington Post,* August 15, 1994, A1.
Gray, Colonel David, USA. Numerous personal interviews by author. 2004–5. Brookings Institution, Washington, DC, and Fort Campbell, KY.
Hackworth, David H. "A Rare Act of Moral Courage." *Defending America,* January 27, 1998. Online magazine. http://www.hackworth.com/27jan98.html.
Halberstam, David. *War in a Time of Peace.* New York: Scribner, 2001.
Halperin, Morton. *Bureaucratic Politics and Foreign Policy.* Washington, DC: Brookings Institution, 1974.
Hann, Captain Chip, USA. Personal interview by author. October 13, 2004. Fort Campbell, KY.
Harris, Art. *Growing up on the Front Line.* CNN.com. March 28, 2003. Accessed October 20, 2003.
Harris, Art. *Marines Recover Bodies of Slain Comrades.* CNN.com, March 28, 2003. Accessed October 2003.
Hayes, Margaret Daly, and Gary F. Weatley, National Institute of Peace Studies. *Interagency and Political-Military Dimensions of Peace Operations: Haiti, a Case Study.* Washington, DC: National Defense University Press, 1996.
Hegland, Corine. "National Security—Why Civilians Instead of Soldiers?" *National Journal,* April 28, 2007.
Herron, Captain David, USMC Reserve. Personal interview by the author. September 22, 2004. Marine Corps Warfighting Lab, Quantico, VA.
Hillen, John. "Playing Politics with the Military." *Wall Street Journal,* December 5, 1996.
Hockstader, Lee. "Army Learns Ropes from City of Austin: 1st Cavalry Prepares to Run Baghdad." *Washington Post,* February 22, 2004, A01.
Holbrooke, Richard. *To End a War.* New York: Random House, 1999.
Holden, Lieutenant Colonel Christopher, USA, Commander, TF Bravo, Joint Readiness Training Center. Personal correspondence, site visit, and numerous interviews by author. March 2004–June 2005. Via telephone and e-mail.
Horvath, Lieutenant Colonel Jan, USA. Personal correspondence and numerous interviews by author. October 2004–June 2005. Telephone and e-mail.
House Committee on Foreign Affairs. *Congressional Testimony of Secretary of State Condoleezza Rice: The New Way Forward in Iraq,* January 11, 2007.

Howard, Michael. "Military Science in an Age of Peace." *RUSI Quarterly* 119, no. 1 (1973): 7.
Huntington, Samuel. *The Soldier and the State: The Theory and Politics of Civil-Military Relations.* 18th ed. Cambridge, MA: Belknap Press of Harvard University Press, 2001.
"Iraq: What Next." Special issue, *New Republic* 2006.
Jaffe, Greg. "As Chaos Mounts in Iraq, U.S. Army Rethinks Its Future: Amid Signs Its Plan Fell Short, Service Sees Benefits of Big Tanks, Translators." *Wall Street Journal,* December 8, 2004.
Jager, Sheila Miyoshi. "On the Uses of Cultural Knowledge." Strategic Studies Institute Published Monograph. Carlisle, PA: U.S. Army War College, 2007. http://www.au.af.mil/au/awc/awcgate/ssi/jager_cultural_knowledge.pdf.
James, Colonel A. "Al" Pace, USMC. "Civil-Military Operations Center." *Marine Corps Gazette* (June 2005).
Jervis, Robert. *Perception and Misperception in International Politics.* Princeton: Princeton University Press, 1976.
Johnson, Lieutenant Colonel Wray. "Whither Aviation Foreign Internal Defense?" *Aerospace Power Journal* (Spring 1997).
Johnson, Wray. *Vietnam and American Doctrine for Small Wars.* Bangkok: White Lotus Press, 2001.
Joint Low Intensity Conflict Project. *Joint Low Intensity Conflict Project Final Report, Executive Summary.* Ft. Monroe, VA: U.S. Army TRADOC, 1986.
Joint Warfighting Center. *Joint Task Force Commander's Handbook for Peace Operations.* Fort Monroe, VA: Department of Defense, June 16, 1997.
Joulwan, General (Ret.) George, USA. Personal interview by author. November 17, 2003. Arlington, VA.
Judis, John B. "Imperial Amnesia." *ForeignPolicy.com* (July–August 2004).
Kaarbo, Juliet, and Ryan K. Beasley. "A Practical Guide to the Comparative Case Study Method in Political Psychology." *Political Psychology* 20, no. 2 (1999): 369–91.
Kagan, Fred. *Choosing Victory: A Plan for Success in Iraq, Phase I Report.* Washington, DC: American Enterprise Institute, 2006.
Kagan, Fred, and Jack Keane. "The Right Type of 'Surge': Any Troop Increase Must Be Large and Lasting." *Washington Post,* December 27, 2007.
Kahl, Colin. "The Four Phases of the U.S. Coin Effort in Iraq." *Small Wars Journal* blog, March 18, 2007. http://smallwarsjournal.com/blog/2007/03/the-four-phases-of-the-us-coin/.
Kahl, Colin. "Shaping the Iraq Inheritance." Washington, DC: Center for a New American Security, June 2008.
Karnow, Stanley. *In Our Image: America's Empire in the Philippines.* New York: Ballantine Books, 1989.
Katz, Ian. "Depressed or Just Decent?" *Guardian,* May 30, 1995, T4.
Kelly, Jack. "Iraq Provides Peacekeeping Institute with Needed Boost." *Pittsburgh Post-Gazette,* November 27, 2003, A-1.
Kelly, Lorelei. "A Military Orphan Faces the Ax." *Boston Globe,* April 26, 2003, Op-ed, A15.
Kier, Elizabeth. "Culture and Military Doctrine: France between the Wars." *International Security* 19, no. 4 (1993).

Kier, Elizabeth. *Imagining War: French and British Military Doctrine between the Wars.* Princeton: Princeton University Press, 1997.

Kilcullen, David. "Dinosaurs versus Mammals: Insurgent and Counterinsurgent Adaptation in Iraq, 2007." Presentation to RAND Insurgency Board, Washington, DC, May 8, 2008.

Kilcullen, David. "Don't Confuse the 'Surge' with the Strategy." *Small Wars Journal* blog, January 19, 2007. http://smallwarsjournal.com/blog/2007/01/dont-confuse-the-surge-with-th/.

Killebrew, Colonel (Ret.) Robert, USA. Personal interview by author. June 10, 2005. Basin Harbor, VT.

Kim, Daniel H. "The Link between Individual and Organizational Learning." *Sloan Management Review* 35, no. 1 (1993): 37–50.

Kitfield, James. *Prodigal Soldiers: How the Generation of Officers Born of Vietnam Revolutionized the American Style of War.* New York: Brassey's, 1997.

Kohn, Richard H. "Out of Control: The Crisis in Civil-Military Relations." *National Interest* 35 (1994): 3–18.

Kolb, D. A. *Experiential Learning.* Englewood Cliffs, NJ: Prentice-Hall, 1984.

Krepinevich, Andrew F., Jr. *The Army in Vietnam.* Baltimore: Johns Hopkins University Press, 1986.

Kretchik, Walter E., Robert F. Baumann, and John T. Fishel. *Invasion, Intervention, "Intervasion": A Concise History of the U.S. Army in Operation Uphold Democracy.* Fort Leavenworth, Kansas: U.S. Army Command and General Staff College Press, 1998. http://www.globalsecurity.org/military/library/report/1998/kretchik.htm.

Krulak, General Charles C., USMC. "The Strategic Corporal: Leadership in the Three Block War." *Marines Magazine* (January 1999).

Krulak, Charles C. "The Three Block War: Fighting in Urban Areas" (speech presented at National Press Club, Washington, DC). *Vital Speeches of the Day*, December 15, 1997, 139–41.

Krulak, Victor H. *First to Fight: An Inside Look at the Marine Corps.* Annapolis: U.S. Naval Institute, 1984.

Lacquement, Richard. *Shaping American Military Capabilities after the Cold War.* London: Praeger, 2003.

Langewiesche, William. "Peace Is Hell." *Atlantic Monthly* (October 2001): 51–80.

Langley, Lester. *The Banana Wars: United States Intervention in the Caribbean, 1898–1934.* Lexington: University Press of Kentucky, 1985.

Lebovic, James H. "How Organizations Learn: U.S. Government Estimates of Foreign Military Spending." *American Journal of Political Science* 39, no. 4 (1995): 835–63.

Leney, Colonel (USA Ret.) Thomas. Personal interview by author. 2003. UN Foundation, Washington, DC.

Levitt, Barbara, and James G. March. "Organizational Learning." *Annual Review of Sociology* 14 (1988): 319–40.

Levy, Jack S. "Learning and Foreign Policy: Sweeping a Conceptual Minefield." *International Organization* 48, no. 2 (1994): 279–312.

Linn, Brian. *The U.S. Army and Counterinsurgency in the Philippine War, 1899–1902.* Chapel Hill: University of North Carolina Press, 1989.

Lowry, Richard. "What Went Right: How the U.S. Began to Quell the Insurgency in Iraq." *National Review,* May 9, 2005.

Lynch, Lieutenant Colonel Thomas, USA. Personal interview by author. 2003. Brookings Institution, Washington, DC.

Mahon, John K. *History of the Second Seminole War.* Gainesville: University of Florida Press, 1967.

"Maneuver Platoon and Company Stability and Support Operations (Saso): Study Prepared for William M. Steele, LTG, Deputy Commanding General for Combined Arms." Fort Leavenworth, KS, 2001.

March, James, and Johan Olsen. "The New Institutionalism: Organizational Factors in Political Life." *American Journal of Political Science* 78, no. 3 (1984).

March, James G., and Johan P. Olsen. "Organizational Learning and the Ambiguity of the Past." In *Ambiguity and Choice in Organizations,* edited by James C. March and Johan P. Olsen. Bergen: Universitets forlaget, 1979.

Marquis, Susan L. *Unconventional Warfare: Rebuilding US Special Operations Forces.* Washington, DC: Brookings Institution Press, 2003.

Mattis, Lieutenant General James, USMC. Personal interview by author. June 15, 2005. Quantico, VA.

Mendelson-Forman, Johanna, and Rick Barton. "The Nation-Building Trap: Haiti after Aristide." *Open Democracy On-Line,* March 11, 2004.

Mentis, Captain Alex, USA. Personal interview by author. October 13, 2004. Fort Campbell, KY.

Miller, Laura L. "Do Soldiers Hate Peacekeeping? The Case of Preventative Diplomacy Operations in Macedonia." *Armed Forces & Society* 23, no. 3 (1997): 415–46.

Millett, Allan R. *Semper Fidelis: The History of the United States Marine Corps.* New York: Free Press, 1991.

Millett, Allan R., and Peter Maslowski. *For the Common Defense: A Military History of the United States of America.* New York: Free Press, 1994.

Mordica, George. Numerous personal interviews by author. March 2004–May 2005. Center for Army Lessons Learned, Fort Leavenworth, KS, and via telephone.

Murdock, Clark, Michele Flournoy, and Chris Williams. *Beyond Goldwater-Nichols: Defense Reform for a New Strategic Era.* Washington, DC: Center for Strategic and International Studies, March 2004.

Nagl, John. Foreword to University of Chicago Press edition. *The U.S. Army/Marine Corps Counterinsurgency Field Manual.* Chicago: University of Chicago Press, 2007.

Nagl, John A. *Counterinsurgency Lessons from Malaya and Vietnam: Learning to Eat Soup with a Knife.* Westport, CT: Praeger, 2002.

Nash, Major General Bill, and John Hillen. "Can Soldiers Be Peacekeepers and Warriors?" *NATO Review,* 49, no. 2 (2001): 16–20. Online edition.

Nash, General William. *FRAGO 136, Commander's Intent.* January 1996.

Nash, General (Ret.) William, USA. Personal interview by the author. October 2003. Council on Foreign Relations, Washington, DC.

NBC News. *Meet the Press.* September 14, 2003. http://msnbc.msn.com/id/3080244/default.htm.

Newbold, Lieutenant General (Ret.) Gregory, USMC. Numerous personal interviews by author. June–July 2005. Arlington, VA.

Newett, Sandra, Anne Dixon, Mark Geis, Linda Keefer, Ken LaMon, Cori Rattleman, Adam Siegel, and Karen Smith. *Emerald Express '95: Analysis Report*. Alexandria, VA: Center for Naval Analysis, 1996.

North, Douglass C. *Institutions, Institutional Change, and Economic Performance*. New York: Cambridge University Press, 1990.

Odierno, Lieutenant General Ray, Commander, Multi-National Corps, Iraq. "Counterinsurgency Guidance." Letter issued to Coalition troops in Baghdad: Multinational Corps Iraq (MNC-I), June 16, 2007.

O'Hanlon, Michael. "The Need to Increase the Size of the Deployable Army." *Parameters* (Autumn 2004): 4–17.

O'Hanlon, Michael, and Adriana Lins de Albuquerque (later Jason Campbell). "The Iraq Index." www.brookings.edu/iraqindex (begun in Fall 2003 and updated weekly).

Oliver, Colonel George, U.S. Army, former Director, Army Peacekeeping Institute. Numerous personal interviews by author. October 2003–June 2005. Washington, DC.

Oliver, Colonel (Ret.) George, U.S. Army, former Director, Army Peacekeeping Institute. "U.S. Peace Operations since 1981: Troop Strengths." Washington, DC: Center for Military History, 2005.

Orr, Robert C., ed. *Winning the Peace: An American Strategy for Post-Conflict Reconstruction*. Significant Issues Series, Center for Strategic and International Studies. Washington, DC: CSIS Press, 2004.

Owens, Lieutenant Colonel Chris, USMC, Director, Marine Corps School of Advanced Warfighting (SAW). Personal interview by author. October 2005. Quantico, VA.

Packer, George. *Assassin's Gate: America in Iraq*. New York: Farrar, Straus and Giroux, 2005.

Packer, George. "The Lesson of Tal Afar." *New Yorker*, April 10, 2006.

Parker, Lieutenant Colonel James, USA. "Some Random Notes on the Fighting in the Philippines." *Journal of the Military Service of the United States* 27 (June 1900): 317–40.

Parton, James. *General Butler in New Orleans*. Boston: Ticknor & Fields, 1866.

Perito, Robert. "Provincial Reconstruction Teams in Iraq." U.S. Institute of Peace briefing. February 2007. http://www.usip.org/pubs/usipeace_briefings/2007/0220_prt_iraq.html.

Perry, William J., Secretary of Defense. Speech given to the Naval Senior Enlisted Academy, Newport, RI. http://www.defenselink.mil/speeches/1996/s19960710-perry.html (accessed May 15, 2005, and July 10, 1996).

Petraeus, Lieutenant General David, USA. "Presentation on Iraq Reconstruction." Washington, DC, 2004.

Platt Amendment. http://ourdocuments.gov/doc.php?doc=55.

Posen, Barry R. "Explaining Military Doctrine." In *The Use of Force*, edited by Robert Art and Kenneth Waltz. Lanham, MD: Rowman and Littlefield, 1999.

Posen, Barry R. *The Sources of Military Doctrine: France, Britain, and Germany between the World Wars*. Ithaca: Cornell University Press, 1984.

Powell, Colin. "News Briefing: 1993 Report on the Roles, Missions and Functions of the Armed Forces, Submitted by Colin Powell, Chairman of the Joint Chiefs of Staff." Washington, DC: Office of the Assistant Secretary of Defense (Public Affairs), February 12, 1993. http://www.fas.org/man/docs/corm93/brief.htm.

Powell, Colin. "U.S. Forces: The Challenges Ahead." *Foreign Affairs* (Winter (2002–3)).

Powell, Colin. "Why Generals Get Nervous." *New York Times,* October 8, 1992, A35.

Powell, Colin, William Odom, John Lehman, Samuel Huntington, and Richard Kohn. "An Exchange on Civil-Military Affairs." *National Interest* 36 (1994): 23–31.

Powell, Colin, and Joseph E. Persico. *My American Journey.* New York: Random House, 1995.

Rice, Condoleezza. "Promoting the National Interest." *Foreign Affairs* 79, no. 1 (2000): 45.

Rice, Susan, former Director of International Organizations and Peacekeeping, National Security Council (1993–95), and former Assistant Secretary of State for African Affairs (1997–2001). Personal interview by author. 2004. Brookings Institution, Washington, DC.

Ricks, Thomas. *Fiasco: The American Military Adventure in Iraq.* New York: Penguin Press, 2006.

Ricks, Thomas. "Marines to Offer New Tactics in Iraq, Reduced Use of Force Planned after Takeover from Army." *Washington Post,* January 7, 2004.

Ricks, Thomas. "Officers with PhDs Advising War Effort." *Washington Post,* February 5, 2007, A01.

Ricks, Thomas. "Politics Collide with Iraq Realities: Commanders Seek Longer-Term Focus." *Washington Post,* April 8, 2007, A01.

Ricks, Thomas. "The Price of PowerGround Zero: Military Must Change for the 21st Century—the Question Is How." *Wall Street Journal,* November 12, 1999, A1.

Ricks, Thomas. "The Widening Gap between the Military and Society." *Atlantic Monthly* (July 1997).

Ricks, Thomas E., and Anne Marie Squeo. "The Price of Power—Sticking to Its Guns: Why the Pentagon Is Often Slow to Pursue Promising Weapons." *Wall Street Journal,* October 12, 1999, A1.

Robbins, Carla Anne. "The Price of Power—Ultimate Threat: U.S. Nuclear Arsenal Is Poised for War—Is It the Right One?" *Wall Street Journal,* October 15, 1999, A1.

Rosati, Jerel. "Developing a Systematic Decision-Making Framework: Bureaucratic Politics in Perspective." *World Politics* 33 (January 1981): 234–52.

Rose, Donald G. "FM 3-0 Operations: The Effect of Humanitarian Operations on U.S. Army Doctrine." *Small Wars and Insurgencies* 13, no. 1 (2002): 57–82.

Rosen, Stephen Peter. *Winning the Next War: Innovation and the Modern Military.* Ithaca: Cornell University Press, 1991.

Salmoni, Barak, Director, Marine Corps Center for Advanced Operational Cultural Learning. Presentation to Marine Corps conference, "Irregular Warfare: Creating Stability in an Unstable World," Center for Emerging Threats and Opportunities, Quantico, VA, July 12, 2005.

Sandler, Stanley. *Glad to See Them Come, Sad to See Them Go: A History of U.S. Army Civil Affairs and Military Government*. S. Sandler, 1994.

Saul, Colonel Larry, USA, Director, Center for Army Lessons Learned. Personal interview by author. March 8, 2004. Fort Leavenworth, KS.

Scales, General (Ret.) Robert, USA. *Certain Victory: The U.S. Army in the Gulf War*. Washington, DC: Potomac Books, 1998.

Schoomaker, General Peter, USA, Chief of Staff. Speech, Center for Strategic and International Studies, Washington, DC, May 15, 2004.

Schultz, Tammy. "Ten Years Each Week: The Warrior's Transformation to Win the Peace." PhD dissertation, Georgetown University, 2005.

Senge, Peter M. *The Fifth Discipline: The Art and Practice of the Learning Organization*. New York: Currency Doubleday, 1990.

Sepp, Kalev. "Best Practices in Counterinsurgency." *Military Review* (May–June 2005).

Sewall, Sarah. Introduction to University of Chicago Press edition. *The U.S. Army/Marine Corps Counterinsurgency Field Manual*. Chicago: University of Chicago Press, 2007.

Shalikashvili, General John M., USA, Chairman of the Joint Chiefs of Staff. "America's Armed Forces: A Perspective." Prepared remarks at the Council on Foreign Relations, New York, November 7, 1996. http://www.defenselink.mil/speeches/1996/t19961107-shali.html.

Shalikashvili, John M. "National Military Strategy: Shape, Respond, Prepare Now: A Military Strategy for a New Era." Washington, DC: Department of Defense, 1997.

Shalikashvili, General John M., USA, Chairman of the Joint Chiefs of Staff. Speech given at Care 50th Anniversary Symposium, Washington, DC, May 10, 1996. http://www.defenselink.mil/speeches/1996/s19960510-shali.html.

Shalikashvili, John M. Speech given to the Robert R. McCormick Tribune Foundation, George Washington University, Washington, DC, May 4, 1995. http://www.defenselink.mil/speeches/1995/s19950504-shali.html.

Shanker, Thom. "Wolfowitz Testifies Pentagon Misjudged the Strength of Iraqi Insurgency." *New York Times*, June 23, 2004.

Smith, R. Jeffrey, and Julia Preston. "United States Plans a Wider Role in U.N. Peace Keeping." *Washington Post*, June 18, 1993, A1.

Snyder, Jack L. *Ideology of the Offensive: Military Decision-Making and the Disasters of 1914*. Ithaca: Cornell University Press, 1984.

Sonsalla, Captain Scott, USA. Address given at Army War College conference, Carlisle, PA, March 13, 2004.

Sorley, Lewis. *A Better War: The Unexamined Victories and Final Tragedy of America's Last Years in Vietnam*. New York: Harcourt, 1999.

"Statement by the Press Secretary: President Clinton Signs New Peacekeeping Policy." White House, May 6, 1994.

Stroup, Theodore G., Jr. "Leadership and Organizational Culture: Actions Speak Louder than Words." *Military Review* 171, no. 1 (1996).

Suarez, Ray. "Gen. Casey Faces Criticism in Senate Confirmation Hearing." *News Hour*, PBS, February 1, 2007. http://www.pbs.org/newshour/bb/military/jan-june07/casey_02-01.html.

Sullivan, General Gordon, USA. Personal interview by author. February 25, 2005. Arlington, VA.

Sullivan, Gordon, and Michael V. Harper. *Hope Is Not a Method: What Business Leaders Can Learn from America's Army*. New York: Random House, 1996.

Summers, Harry H., Jr. *On Strategy: The Vietnam War in Context*. Carlisle Barracks, PA: U.S. Army Strategic Studies Institute, 1982.

Summers, Harry H., Jr. "A War Is a War Is a War Is a War." In *Low Intensity Conflict: The Pattern of Warfare in the Modern World*, edited by Loren B. Thompson. Lexington, MA: Lexington Books, 1989.

Swan, Guy. "Swan on Swain." *Military Review* 5 (May 1988): 86.

Swannack, Colonel Charles H., U.S. Army, and Lieutenant Colonel David R. Gray, U.S. Army. "Peace Enforcement Operations." *Military Review* 76, no. 6 (November–December 1997): 3–10.

Sweeney, Lieutenant Colonel Patrick, USA, Lieutenant Colonel Charles Eassa, USA, Lieutenant Colonel Trent Cuthbert, USA, and Captain Justin Mufalli, USA. *CALL Newsletter: Targeting for Victory: Winning the Civil Military Operations* (September 2003).

Taw, Jennifer Morrison. *Operation Just Cause: Lessons for Operations Other than War*. Santa Monica, CA: RAND Corporation, Arroyo Center, 1996.

Toolan, Colonel John, USMC, Director, Marine Corps Command and Staff College. Presentation to Marine Corps conference, "Irregular Warfare: Creating Stability in an Unstable World," Center for Emerging Threats and Opportunities, Quantico, VA, July 12, 2005.

Trebilcock, Major Craig, USA. "The Myth of Posse Comitatus." *Journal of Homeland Security* (October 2000).

Triggs, Staff Sergeant Marcia. "Fighting in Urban Terrain Challenging, Not Impossible." *Army News Service*, November 27, 2002.

Ucko, David. "Innovation or Inertia: The U.S. Military and the Learning of Counterinsurgency." *Orbis* (Spring 2008).

USMC. *FMFM-1 Warfighting*. Washington, DC: Department of the Navy, 1989.

USMC. *Small Wars Manual of the United States Marine Corps*. Washington, DC: U.S. Government Printing Office, 1940.

Van Evera, Steven. "The Cult of the Offensive and the Origins of the First World War." *International Security* 9, no. 1 (1984): 58–107.

Van Riper, Lieutenant General (Ret.) Paul. Personal interview by author. January 15, 2006. Arlington, VA.

Van Riper, Major Stephen, USMC, 1st Marine Division Director of Training. Personal interview by author. March 9, 2004. Camp Pendleton, CA.

Vector Research. "The 21st Century Army: Roles, Missions and Functions in an Age of Information and Uncertainty." Ann Arbor, MI, 1995.

Vetock, Dennis J. *Lessons Learned: A History of US Army Lesson Learning*. Carlisle Barracks, PA: U.S. Army Military History Institute, 1988.

Viotti, Paul R. "Introduction: Military Doctrine." In *Comparative Defense Policy*, edited by Frank B. Horton III, Anthony C. Rogerson, and Edward L. Warner III. Baltimore: Johns Hopkins University Press, 1974.

Walters, Captain Steve, USA. Personal interview by author. October 13, 2004. Fort Campbell, KY.

Warner, Brigadier General Volney J., USA, Deputy Commandant, Command and

General Staff College. "Evolving and Adapting to Meet Challenges and Exploit Opportunities: A Presentation for Ms. Tammy Schultz." Fort Leavenworth, KS, April 2005.

Washington, George. "Sentiments on a Peace Establishment." 1783. http://www.potowmack.org/washsent.html.

Watkins, David, doctrine developer. Personal interview by author. February 7, 2005. Joint and Allied Doctrine Division, Requirements Integration Directorate, U.S. Army Training and Education Command.

Weigley, Russell F. *History of the United States Army.* New York: Macmillan, 1967.

Wenger, Etienne. *Communities of Practice: Learning, Meaning, and Identity.* New York: Cambridge University Press, 1998.

Wert, Jeffery D. *Mosby's Rangers: The True Adventures of the Most Famous Command of the Civil War.* New York: Simon and Schuster, 1990.

West, Bing. *The Village.* New York: Pocket Books, Simon and Schuster, 1972.

White, Major John C., USMC. "American Military Strategy during the Second Seminole War." Master's thesis, Military Studies, Marine Corps Command and Staff College, 1995.

Wickham, General John. Interview with Chief of Staff of the Army General John Wickham. *Armed Forces Journal* (September 1985). Reprinted in General John A. Wickham, Jr., Army Chief of Staff, *Collected Works of the Thirteenth Chief of Staff of the United States Army* (1987), 342.

Wickham, John. Interview with Chief of Staff of the Army General John Wickham. *Army* (September 1986). Reprinted in General John A. Wickham, Jr., Army Chief of Staff, *Collected Works of the Thirteenth Chief of Staff of the United States Army* (1987), 350.

Wickham, General John A., Jr., Army Chief of Staff. "Introductory Letter: White Paper 1984, Light Infantry Division." *Collected Works of the Thirteenth Chief of Staff of the United States Army* (1987): 311.

Wilkie, Brigadier General Dennis A. "Deja Vu: 'Who's in Charge?'" *Scroll and Sword, Journal and Newsletter of the Civil Affairs Association* 56, no. 3 (2003): 12–15.

Williams, Cynthia. *Filling the Ranks: Transforming the US Military System.* Cambridge, MA: MIT Press, 2004.

Williams, Colonel Garland "Winky." *Executive Summary: Engineering Peace—The Role of the Military in Post Conflict Reconstruction.* Washington, DC: U.S. Institute of Peace, 2005.

Williams, Colonel Garland "Winky." Lecture, "The Military Role in Post Conflict Reconstruction," Washington, DC, May 2, 2005.

Wilson, Major Isaiah, USA. "Educating the Post-Modern U.S. Army Strategic Planner: Improving the Organizational Construct." United States Command and General Staff College, School of Advanced Military Studies, 2003.

Wilson, Major Isaiah, USA. "Thinking beyond War: Civil-Military Operational Planning in Northern Iraq." Paper presented at 2004 Annual ISAC/ISSS conference Washington, DC, September 14, 2004.

Wilson, James Q. *Bureaucracy.* New York: Basic Books, 1989.

Wingenbach, Lieutenant Colonel Karl, doctrine developer. Personal interview by author. February 7, 2005. Joint and Allied Doctrine Division, Requirements Integration Directorate, U.S. Army Training and Education Command.

Wong, Leonard. *Developing Adaptive Leaders: The Crucible Experience of Operation Iraqi Freedom*. Carlisle, PA: U.S. Army War College, Strategic Studies Institute, July 2004.

Wright, Donald P., Colonel Timothy R. Reese, and the Contemporary Operations Study Team. *On Point II: Transition to the New Campaign—The United States Army in Operation Iraqi Freedom, May 2003–January 2005*. Combat Studies Institute Press, U.S. Army Combined Arms Center, 2008.

Ziemke, Earl F. *The U.S. Army in the Occupation of Germany, 1944–1946*. Army Historical Series. Washington, DC: Center of Military History, U.S. Army, 1990.

Zinni, General (Ret.) Anthony, USMC. "How Do We Overhaul the Nation's Defense to Win the Next War?" Address to the Marine Corps Association and U.S. Naval Institute Forum, Arlington, VA, September 4, 2003.

Zinni, General (Ret.) Anthony, USMC. Personal interview by author. January 31, 2005. Arlington, VA.

Zisk, Kimberly Marten. *Engaging the Enemy: Organization Theory and Soviet Military Innovation 1955–1991*. Princeton: Princeton University Press, 1993.

Index

Note: Page numbers that are italicized indicate figures or tables.

Abizaid, John, 152, 173–74
Adelman, Ken, 163
Administration of Civil Affairs in Occupied Alien Territory, 58, *59*
Afghanistan: reconstruction operations in, 126–27; tactical lessons learned fed to training centers from, 196–97; veterans of, on PME curricula relevance after, 156
after-action reviews (AARs): from the Balkans, troop training and, 120; Center for Army Lessons Learned as depository for, 104, *105*; experiential learning and, 110–12; flight training programs and, 100; in Haiti, 102, 170; JRTC's use of, 117; National Training Center's use of, 101–2; in Panama, 136; post–Vietnam War learning culture of U.S. Army and, 21; U.S. Army's success in Iraq due to use of, 189; use in today's military, 97
Agency for International Development. *See* U.S. Agency for International Development
Agoglia, John, 150–52, 164
Agriculture, U.S. Department of, 166
Aideed, Mohammad Farrah, 80
Air Assault (Army light battalion), 99, 116
Airborne (Army light battalion), 116
Air Force, U.S., 99–100, 151. *See also* U.S. military
AirLand Battle Doctrine: education and training to support, 195; FM 100-5 (*Operations,* 1986) focus on, 134; introduction of, 139; learning through doing

and, 195–96; military change after Vietnam War and, 130; peace operations doctrine resisted by leadership in favor of, 157; small wars, counterinsurgency and, 114
Albright, Madeline, 80, 81–82, 122, 168
Allison, Graham, 11
Al Qaeda in Iraq (AQI), 174, 180, 181
Alywin-Foster, Nigel R. F., 188–89, 188n
American Enterprise Institute (AEI), 175, 176–77
American Expeditionary Force (AEF), in World War II, 54, 55–56
"American Military Government of Occupied Germany, 1918–1920" (Hunt, March 4, 1920), 54. *See also* Hunt Report
American public opinion: debate on soldier's roles in military governments and, 61–64; on harsh tactics in the Philippines, 41; of the Vietnam generation, PPD-25 and, 84
amphibious warfare, 52–53, 157, 194
Angola, post–Cold War peacekeeping operations in, 70
Application of Peace Enforcement (PE) at the Brigade and Battalion Level, The, 123, 141
Army, British, Nagl on organizational learning by U.S. Army vs., 20, 192. *See also* British military
Army, U.S.: Alywin-Foster on modern culture vs. Vietnam-era culture of, 188–89; Avant on military culture and adaptation by, 17–18; big-war paradigm of, 73–75, 77; First Armored Division predeploy-

227

Army, U.S. (*continued*)
    ment exercises of, 119–20; frontier duties for 19th-century soldiers of, 29–30; history of MOOTW experiences by, 10, 191; on informal networks, 198; on infrastructure reconstruction, 51–52; as Modular Force, 109–10; Nagl on organizational learning by British Army vs., 20, 192; NGOs' interactions with, 121; Philippines atrocities by, 44–45; political resistance to learning in, 53, 65; post–Cold War irregular warfare and leadership development in, 71–72; post–Vietnam War learning mechanisms, 97; schools for professional military education of, 154–56; use of "small wars" by, 207. *See also* U.S. military; U.S. Military Academy
Army–Air Force Center for Low Intensity Conflict, Langley, Va., 134
Army Infantry School, U.S. (USAIS), 123
Army in Vietnam, The (Krepinevich), 75
Army of Excellence initiative, 115
"Army Training: National Training Center's Potential Has Not Been Realized" (GAO, 1986), 103, 104
Army War College: on cultural knowledge in counterinsurgency, 184; G1 (personnel) War Plans Division planning course at, 58; Historical Section compiling World War I record at, 57; Peacekeeping Institute (PKI) at, 119, 148–52, 168, 189, 195; resistance to change educational curricula of, 97–98
Arnold, Steven L., 80, 88–89, 185
Art of War, The (Jomini, 1838), 29
atrocities. *See* torture
Avant, Deborah, 16–18

Bailey, Ronald, 1, 7
Baird, Lloyd, 102
Baker-Hamilton Commission, *Iraq Study Group Report* and, 177
Balkans: doctrine-writing after operational lessons learned in, 133; learning through doing in, 195; lessons learned from, 126; light forces previously deployed to Somalia in, 116–17; mission rehearsal exercises for, 196; post–Cold War length of mission in, 72; post–Cold War peacekeeping operations in, 70; rules of engagement for peacekeeping in, 91; training for Dayton Peace Accords enforcement in, 119–20; U.S. political distaste for nation building in, 92
Banana Wars (1914–1934): generational split between writers and sponsors of doctrine in, 157; Marine Corps's pacification duties during, 45; Marine Corps tactics, techniques, and politics in, 46–50; as small wars, 207
Base Force report, Bush administration, 86n
Basic Manual for Military Government by United States Force, 58, 59
Bell, J. Franklin, 42
belligerents, JRTC training for separation of, 120
Bensen, Kevin, 164–66
Berlin Wall, dismantling of Cold War and, 67
Bickel, Keith B., 44, 45, 46
Biddle, Steve, 178–79
big war. *See* war, big or major
Birtle, Andrew J., 36
Blue Force Tracker, 117
Board of Economic Welfare, Allied occupation in North Africa and, 63
Boer Wars, 17, 207
Bolger, Daniel P., 112–13
Border Patrol, U.S., *posse comitatus* and, 36
Bosnia: challenges for First Armored Division in, 119–20; criticism of U.S. military in, 93–94; JRTC training for, 123; Nash on enforcing Dayton Accords in, 92–93; Nash on lessons learned from deployment in, 88; post–Cold War length of mission in, 72; Shalikashvili on MOOTW and, 143; U.S. military and new world order in, 67–68. *See also* Balkans
bottom-up learning: in Iraq, 173, 175; military structure, culture, and politics and, 6; Nagl's organizational model of, 160; post–Cold War, experience-driven, 196, 197–98, *197*; success in stabilization and reconstruction operations and, 185. *See also* experiential learning
Bottom-up Review, Clinton's (1993), 85, 86n
Breckenridge, James, 50
Bremer, Paul, 165, 187, 190
brigade combat team (BCT), as Army's basic deployable unit, 166
Brinkley, Paul, 187

British military, Avant on military adaptations by, 17, 18. *See also* Army, British
Brooke, John R., 42
Brookings Institution, 175, 176
budgets, military: military change and, 11–13; post–Cold War reductions of, 69
Builder, Carl H., 14, 15
Bureaucracy Does Its Thing (Komer), 75–76
bureaucratic politics, military change and, 13–14
Bush, George H. W., 68, 91
Bush, George W., and administration: announces Iraq Surge, 159–60; on Powell Doctrine Vietnam syndrome, 79; U.S. military organizational culture at start of presidency by, 90
Butler, Smedley D., 48–49
Byrnes, Kevin, 111–12, 124n, 158

Cagle, Brian, 112–13
Caldwell, William, 132, 152, 153
CALL. *See* Center for Army Lessons Learned
CALL-COM software, 107
CALL Newsletter, 108
Callwell, C. E., 49, 207
career structures: military change through civilian interventions regarding, 17n30; Rosen on military change by maverick officers and, 17
Casey, William, 173–74, 182
Cassidy, Robert C., 73, 77, 83
casualties: aversion to, in PPD-25, 84; fear of, peacekeeping operations and, 91, 92–93; minimization in Vietnam with heavy firepower, 74–75; Powell Doctrine on avoidance of, 78; 2-MTW strategy on, 85; during U.S. Civil War, 34; Vietmalia syndrome as aversion to, 90
Cavalry Drill Regulations (1914), 39, 43–44
Center for Army Doctrine Development, 151
Center for Army Lessons Learned (CALL): adaptations in 1990s to process facilitated by, 123; bottom-up experiential learning leveraged at, 97; combat training centers and, 195; on CompanyCommand.com, 124n; data collection, analysis, and dissemination by, 105–7; expansion from training to real-life operations, 104–5; exposure to MOOTW through, 199; impetus for, 102–4; JRTC development and, 113; lessons learned posted on website of, 198; *Operations other than War*, vol. 4, *Peace Operations* (pamphlet) of, 123; organizational learning theory applied at, 21; PKI absorption into, 149–50; trends analyses, dissemination, and customer service, 108–10. *See also* military learning
Center for the Advancement of Leader Development and Organizational Learning, 124
Central Command (CENTCOM) for Iraq, 164, 190
Central Intelligence Agency (CIA), 167
CERP (Commanders Emergency Response Program), 187
CGSC. *See* Command and General Staff College
Chaffee, Adna R., 40
Cheney, Richard, 162
Chiarelli, Peter, 94–95, 174, 187, 189
Choosing Victory: A Plan for Success in Iraq (Kagan and Keane, 2007), 176–77
civil affairs: current personnel for as reservists, 64; differences between military government and, 57n; institutionalizing military government and, 56–59. *See also* civilian agencies, U.S.
Civil Affairs Division (CAD), of War Department, 64
Civil Affairs Doctrine (1943), 59
Civil Affairs Training Program (CATP), 59, 61
civilian agencies, U.S.: cited in FM 100-23 (*Peace Operations*), 141; CORM on MOOTW and, 87; Emerald Express (civil-military exercise) and, 168–69; JP 3-08 (on interagency operations) and, 137–38; JP 3-08 on responsibility and capability of, 169–70; lack of capacity and capability issues for S&R of, 166–67; lack of learning culture in, stabilizing complex cultures and, 161; lack of personnel to participate in military exercises, 172; Marines and Army in Somalia operation reach out to, 168; military's learning system and, 188; vetting language of FM 3-07, 172–73. *See also* interagency
Civilian Operations Revolutionary Development Support (CORDS), during Vietnam War, 167–68
civilians in the battlefield (COBs), JRTC training using, 117, 120

230 · Index

civil transportation, U.S. Army vs. Marine Corps on reconstruction of, 52. *See also* infrastructure

Civil War, U.S., U.S. Army and rebel insurgents during, 32–34

Clark, Wesley, 103, 113

Clarke, Walter, 121

Clausewitz, Carl von, 9, 24, 29

clear, hold, and build strategy, for Iraq, 171, 173, 174–75

Clinton, Bill, and administration: expanding U.S. role in UN peacekeeping and, 81–82; issues impeding promotion of peace operations by, 83; on military reductions in 1992, 68; National Security Strategy (NSS) on 2-MTW, 85–86; on peacekeeping role of U.S. military, 80–81; PKI importance during, 149; Presidential Decision Directive-25 (PDD-25) of, 141

Coalitional Provisional Authority (CPA), Iraq, 165. *See also* Bremer, Paul

Coast Guard, U.S., *posse comitatus* and, 36

Cold War: attitudes during and lessons of Vietnam, 73–77; breakdown in cultural resistance to MOOTW after, 191–92; dismantling Berlin Wall and, 67; MOOTW doctrine available at end of, 134–38; offensive operations during, 72n. *See also* new world order; post–Cold War bureaucratic politics model

collection teams: CALL system and, 105; four-block war learning system and, 186

collective learning: consensus for organizational learning and, 25–26; generational learning and, 24–25. *See also* military learning

Combat Instructions, two editions for American Expeditionary Force of, 55

Combat Maneuver Training Center (CMTC), 116, 119–20, 196, 201–2

Combat Training Center Program, 97

combat training centers (CTCs): data collection by, 106–7; development and characteristics of, 99–100; doctrine-writing in the 1990s and, 133–34; experiential to organizational learning in, 110–12; exposure to MOOTW through training scenarios at, 199; learning through doing at, 195–96; transformation of Army's training programs and, 126. *See also* Center for Army Lessons Learned

combined arms assessment teams (CAATs): data collection workshops for, 106–7; formation of and mission for, 105–6

Combined Arms Center blog, 183

Command and General Staff College, Army: professional military education at, 154; resistance to change educational curricula of, 97–98; *Small Wars Manual* written for, 50

Command and Staff College, Quantico, Va., 154

Commerce, U.S. Department of, 166

Commission on Roles and Missions (CORM), 86

communities of practice (CoPs): generational learning and, 25; officer commissioning class of 1995 and, 202; in organizational learning theory, 25–26; technology-enabled, 124–26; use for Surge in Iraq, 183

CompanyCommand.com/Company Command.mil, 124, 125–26, 125n, 183, 198

Congress, U.S., readiness concerns of late 1990s of, 89–90

contractors, as civilian agency actors in Iraq, 172

Corps of Engineers, U.S., U.S. Military Academy graduates and, 29

counterinsurgency (COIN): AirLand Battle Doctrine and, 114; as clear, hold, and build strategy in Iraq under Casey, 173, 174; counterterrorism vs., 204; definition of, 4, 203; institutional processes hindering teaching lessons learned from, 65; Light Infantry and Special Forces for, 116; PME relevance for institutional learning on, 156–57; PPD-25 on, 85; understanding of during Surge in Iraq, 181–82; U.S. Army's big-war paradigm and, 74. *See also* FM 3-24; irregular warfare; military operations other than war; stability operations

Counterinsurgency (FM 3-24, 2006), *138*, 152–53, 177–79, 183

Counterinsurgency Academy, Taji, Iraq, 182

"Counterinsurgency Guidance" (Odierno letter), 179, 183, 184

Counterinsurgency Operations (FMI 3-07.22), 152

Counterinsurgency Reader, 183

counterterrorism (CT), 203–4
Craig, Edward, 47
Crane, Conrad, 152
Crocker, Ryan C., 7
CSC. *See* Command and Staff College
CTC. *See* combat training centers
CTC Quarterly Bulletin, 107, 108
Cuba: coping and adapting after Spanish-American War in, 38–41; lessons from the Philippines applied to, 42–45; Marine Corps' view of Army operations in, 47
Cuban Liberation Army, 42
cultural awareness, enhanced, for Surge in Iraq, 183–84
cultural resistance to change, during peacetime vs. wartime, 158. *See also* organizational culture
Cummings, Henry, 60

Davidson, Janine, 188n
Dayton Peace Accords, 92, 120
Defense, U.S. Department of: assistant secretary of defense for special operations and low-intensity conflict and integrating capabilities (ASD for SOLIC/IC), 206; on counterinsurgency, 203; on counterterrorism, 203; on foreign internal defense and internal defense and development, 204; on insurgency, 204; on irregular warfare, 205; on peacekeeping role of U.S. military, 86; on stability operations, 150–51; support for operational capacity of UN peacekeeping operations and, 82; on unconventional warfare, 208. *See also* Pentagon
demographic targeting, for Surge in Iraq, 184, 184n
Dilegge, Dave, 183
dirty little operations/wars, 114, 118
discovery learning, 125, 125n. *See also* experiential learning
doctrinal manuals: on duties during occupation, 58, *59;* lack of, for early JRTC training exercises, 122–23. *See also* handbooks; *under* FM; *under* JP
doctrine: bottom-up organizational learning model and revisions to, 160; on capability and capacity of civilian agencies for S&R, 167; changing role in military learning cycle of education and, 129–32; for COIN and stability operations, lack of consensus on, 187; development of for the new world order, 138–47, *138;* education, the training revolution and, 195–200; and education for operations other than war, 129; Emerald Express ideas prematurely fed into, 169; FM 3-0 (*Operations*) on, 129–30; formal and informal, for Surge in Iraq, 183; model for development of, 130–31; on MOOTW and peace operations, 1990s writing of, 132–34; PKI and next generation of, 148–52; Sullivan's top-down revision to, 131–32; top-down, traditional development of, 157
Dodge, Toby, 178–79
Dominican Republic: lessons from Haiti applied to Marine Corps in, 47–48; Marine Corps' pacification duties in (1914–1934), 45; Marine Corps tactics, techniques, and politics in, 46; military training center in, 49; political constraints on Marines in, 51
Downie, Richard D., 14, 19, 76n
Drug Enforcement Agency, 36
drug war: *posse comitatus* and, 36; in Western Hemisphere, Joulwan and, 140
Dumas, Wess, 109n

East Timor, post–Cold War peacekeeping operations in, 70
ECLIPSE II, 164–65
economic development, four-block war and, 186
Ecuador, post–Cold War peacekeeping operations in, 70
education: changing role in military learning cycle of doctrine and, 129–32; doctrine, the training revolution and, 195–200; and doctrine for operations other than war, 129; as driver of change in post–Vietnam era, 155–56. *See also* professional military education; schools, military; war colleges
82nd Airborne: JRTC scenarios and lessons learned in Panama by, 116–17; Operation Just Cause in Panama (1989) and, 67; peace operations exercise at JRTC (1994) for, 122
Eisenhower, Dwight D., 63–64
Ellis, E. H., 49, 50
Emerald Express (Marine Corps exercise), 168–69

engagement: nonmilitary actors and restricted rules of, 72; post–Cold War operations and length of, 70, 72; rules of, as force protection, 93; rules of, peacekeeping in Haiti and the Balkans and, 91; rules of, Weinberger-Powell Doctrine on, 78

ethnic strife: in Iraq through repression of Shia majority, 162; JRTC training for, 120

European military theory: Napoleonic Wars as basis for, 29; used in U.S. military schools, 36

exit strategies: Powell Doctrine and PPD-25 on, 84; Weinberger-Powell Doctrine on, 78

experiential learning: Byrnes on peacetime vs. wartime variations in, 158; Kolb's cycle of, 110, *111;* lack of structures for capturing and analyzing, 37; officer commissioning class of 1995 and, 202; in organizational learning theory, 24; organizational structures and process to promote, 192–93; post–Vietnam War system for capturing, 195; in Somalia, Haiti, and the Balkans on peace operations, 163. *See also* bottom-up learning

external actors or pressure: military change forced by, 16; modern Army's propensity to learn and, 193; organizational theory on military change and, 12–13. *See also* nonmilitary actors; political systems

Fastabend, David, 145–46, 147
FC 100-20 (*Operations*, 1986), 135. *See also* FM 100-20
feedback loop to headquarters, for Surge in Iraq, 182–83
field coaches, for Surge in Iraq, 182–83
field manuals. *See* doctrinal manuals; *under* FM; *under* JP
field regulations, as resources for Philippine and Cuban campaigns, 39–40, *39*, 43–45
Field Service Regulations (1905), 43
Fifth Discipline, The (Senge), 20
First Armored Division, CMTC predeployment exercise for, 119–20
First Gulf War. *See* Persian Gulf War
First Marine Expeditionary Force, 79, 168
Fisher, George A., 120

Flavin, Bill, 151
Fleet Marine Force Manual 1 *Warfighting* (1989), 135
FM 3-0 (*Operations*, 2001), 129–30, *138*, 142
FM 3-0 (*Operations*, 2008), *138*, 144–47
FM 3-07 (*Stability and Support Operations*, 2003), *138*
FM 3-07 (*Stability and Support Operations*, 2008): Army's PKI and development of, 148, 151, 153; Leonard's use of nonmilitary practitioners to vet language for, 172–73; Oliver on FM 3-0 and development of, 147; published by University of Michigan Press, *138;* on supporting civilian government actors, 169
FM 3-07.31 (*Multi-service Manual for TTP for Peace Operations*, 2003), 144, 151
FM 3-24 (*Counterinsurgency*, 2006), *138*, 152–53, 177–79, 183
FM 3-90 (*Tactics*), 147
FM 7-98 (*Operations in a Low-Intensity Conflict*), 123
FM 27-5 (*Military Government and Civil Affairs*, 1940), 59, *59*, 60
FM 27-10 (*Rules of Land Warfare*, 1939), 58, 59, *59*, 60
FM 90-10 (*Urban Operations*) (1979), 134, *135*
FM 100-5 (*Operations*, 1986), 134, *135*
FM 100-5 (*Operations*, 1993): challenges in northern Iraq and Somalia and, 137; on civil-military coordination for future operations, 138; development of, 139–40; difficulties in updating, 197; MOOTW included in, 119, *138;* Sullivan and writing of, 131. *See also* FM 3-0
FM 100-20 (*Military Operations in a Low-Intensity Conflict*), 122–23, 134, *135*, 136
FM 100-20 (*Stability and Support Operations*), 197–98
FM 100-23 (*Peace Operations*, 1994), *138*, 140–41
FMI 3-07.22 (*Counterinsurgency Operations*), 152
fog of war, as obscuring fundamental truths, 157
force, use of: Clinton doctrine on peace operations and, 83; inappropriate, 137n; Weinberger-Powell Doctrine on, 78
Ford, Robert, 178–79

Foreign Affairs, Krepinevich article critical of Bush administration in, 175–76
foreign internal defense (FID), 204
foreign militaries, Emerald Express (civil-military exercise) and, 168–69
foreign service officers (FSOs), number and skill sets of, 166–67
Fort Chaffee, Ark., 116
Fort Irwin, Calif. *See* National Training Center
Fort Leavenworth, Kans., 37, 37n. *See also* Command and General Staff College, Army
Fort Polk, La., 116, 117–18, 119. *See also* Joint Readiness Training Center
forward operating bases (FOBs), in Iraq, 174–75, 179
four-block war, learning system limits and, 185–88, 202
Franks, Fred, 140
Franks, Tommy, 152
Freedmen's Bureau, 35
French military style. *See* Napoleonic Wars
frontier, American: guerrilla war and nation building by 19th century soldiers in, 29–30; irregular operations tactics applied to Civil War from, 32
full-spectrum operations: FM 3-0 (*Operations*, 2008) and, 144–47; FM 100-5 (*Operations*, 1993) and, 197; three-block war compared to, 185
"future concept" (theory), development of, 131

G1 (personnel) War Plans Division planning course, 58, 59
G2 (intelligence) War Plans Division, 60
G3 (operations) planners of General Staff (G1), 58–59, 60
G5 training section, of U.S. Army, 54–56
Gates, John, 31, 43
Gendarmerie Rules (Butler), 49
General Accounting Office (GAO), 89–90, 103, 104
General Order 100 (1863, 1898), 33, 39, 39. *See also* Lieber Code
generational learning: current Army training programs and, 126; military's capacity for institutional learning and, 199–200; in organizational learning theory, 24–25; post–Cold War, officer career comparison in, 200–202; staff colleges as education link in, 154–55; Vietnam War and, 76
Global Positioning System (GPS), National Training Center's use of, 101
Goldwater-Nichols Act (1986), 86, 86n, 115, 133
Gordon, Gary, 80n, 118
Gorman, Paul, 99, 100, 200
governance, four-block war and, 186
Gray, David, 109, 121–23
Grenada, U.S. mission in, 72, 115
guerrilla tactics or warfare: on American frontier, 29–30; Chaffee's use of, 40; definition of, 204; JRTC training for, 118; prevention in Cuba of, 42; unconventional warfare and, 208; during U.S. Civil War, 32. *See also* irregular warfare
Gullion, Allen W., 58, 60, 62, 64

Haiti: after-action review on civilian capacity issues, 170; after-action reviews used in, 102; Army response to hostile mob in, 112–13; doctrine-writing after operational lessons learned in, 133; Gendarmerie, Butler forms school for, 48–49; JP 3-07 (*MOOTW*, 1995) and mission to, 142; JRTC training (1994) for deployment to, 122; learning through doing in, 195; light forces previously deployed to Somalia in, 116–17; Marine Corps' pacification duties in (1914–1934), 45; Marine Corps tactics, techniques, and politics in, 46–50; mission rehearsal exercises for, 196; Operation Uphold Democracy in, 71; political constraints on Marines in, 51; post–Cold War length of mission in, 72; rules of engagement for peacekeeping in, 91; troop types used in, 71–72; U.S. military and new world order in, 67–68; veterans of, on PME curricula relevance after, 156
Halleck, Henry W., 32
Halperin, Morton, 13, 14
handbooks, four-block war learning system and, 186. *See also* doctrinal manuals; *under* FM; *under* JP
Hann, Chip, 109n
Harvey, Derek, 178–79
Hilldring, John H., 59, 64
Hillen, John, 88
Holbrooke, Richard, 79, 84, 92, 119
Horvath, Jan, 152

humanitarian assistance, JRTC training for, 120
humanitarian goals, post–Cold War operations and, 70–71
Hunt, Irvin L., 53–54, 57
Huntington, Samuel, 43
Hunt Report (1920), 54, 59
Hussein, Saddam, 78–79, 162

IDAD (Internal Defense and Development), 204–5
IFOR. *See* Bosnia
individuals: generational learning and, 24–25; in informal networks and communities of practice, 25–26; isolation during peacekeeping operations, rules of engagement and, 93–94; organizational learning and, 19; role in military learning of, 190; rotation system during Vietnam War for, 77; as visionaries for PKI and doctrine development, 151–52
Infantry and Drill Regulations (1898), 39, 40
Infantry and Drill Regulations (1911), 43–44
infantry school, Army's, incorporating MOOTW into doctrinal manuals and, 151
informal learning, in organizational learning theory, 24. *See also* experiential learning
informal networks: bottom-up learning and, 198; conscious effort to accelerate, 196; generational learning and, 25; in organizational learning theory, 25–26; Vietnam War and, 76. *See also* communities of practice
infrastructure: in post-invasion Iraq, U.S. military planners on, 165; three-block war and, 186; U.S. Army vs. Marine Corps on reconstruction of, 51–52
inspector-instructors, of Army's G5 training section, 54–55
institutional learning cycle: collecting and sharing in, 156; phases in, 175. *See also* organizational learning
institutional memory: informal transfer in U.S. Marine Corps of, 20–21; organizational culture and, 14; organizational learning and, 19
insurgency/insurgent forces: definition of, 204; JRTC training for interactions with, 117–18, 120; U.S. military planners on post-invasion Iraq risk of, 165

interagency, capability and capacity of U.S. military vs., 166. *See also* civilian agencies
intergovernmental organizations: cited in FM 100-23 (*Peace Operations*), 141; post–Cold War operations and, 71. *See also* nongovernmental organizations
Internal Defense and Development (IDAD), 204–5
internally displaced persons, post–Cold War operations and, 71
International Law (Halleck, 1861), 39–40, 39
Internet sites. *See* websites
interpretation, in the learning cycle, 22, 23
in-theater training, for Surge in Iraq, 182
Introduction to the Study of International Law (Woolsey, 1864), 39
Iran: failure to rescue hostages at Desert One in, 115; Iraq's strained relations with before U.S. invasion of Iraq, 162
Iraq: adequacy of FM 100-5 and JP 3-0 for challenges in, 137; adequacy of LIC manual on challenges in, 136; AirLand Battle Doctrine and, 130; American military culture and planning for, 163–66; Bush announces Surge in, 159–60; counterinsurgency and U.S. military training adaptations, 123; doctrinal support for Operation Provide Comfort in, 134; failures in strategic and political leadership for operation in, 161–66; invasion of, as strategic judgment error, 161–63; on-the-job "discovery learning" in, 125; learning to Surge in, 173–85; Marines take control of Nasiriya (2003), 1; misperceptions about capacity and capability of civilian agencies in, 170–71; myth of interagency support for, 166–73; peace operations in, CORM on, 87–88; Petraeus's stability and reconstruction plan for, 94–95; room for improvement for reconstruction operations in, 126–27; tactical lessons learned fed to training centers from, 196–97; unemployed men with occupying army in, 42; U.S. military and new world order in, 67–68; veterans of, on PME curricula relevance after, 156. *See also* Surge for Iraq
Iraq Planning Group study, AEI's, 176–77
Iraq Study Group Report (Baker-Hamilton Commission), 177

irregular warfare (IW): Cold War vs. post–Cold War, 71–72; definition of, 205; JRTC training for, 118; unconventional warfare and, 208; U.S. Army's big-war paradigm and future enemies' use of, 75; during U.S. Civil War and Reconstruction, 32–34; use of term, 5; Vietnam War lessons for, 77. *See also* guerrilla tactics or warfare

Israel, CALL team observations in, 108

Jefferson, Thomas, 29
Johnson, Andrew, 35
Joint Center for Lessons Learned, 105
Joint Chiefs of Staff: joint operating concept of Joint Forces Command and, 131; JRTC support by, 122; National Military Strategy and, 86n. *See also* Pace, Peter; Powell, Colin; Shalikashvili, John
Joint Forces Command, 131
Joint Knowledge Online, 198
joint operating concept (JOC), 131
joint publications. *See under* JP
Joint Readiness Training Center (JRTC): combat training centers and development of, 99; experiential learning at, 117–18; impetus for, 113–14; Light Infantry and Special Operations, 114–15; mission rehearsal exercises at, 196; original focus on MOOTW for, 118–23; results of training at, 112–13; Shugart-Gordon (village) in, 8on
Joint Task Force Commander's Handbook for Peace Operations (1999), *138*
Jomini, Antoine Henri de, 29, 193
Joulwan, George, 140
journal articles: Army-sponsored, for sharing military experiences, 39; Army-sponsored, lessons of the past in, 40; Marine Corps lessons learned in, 48; military learning through, 97. *See also CTC Quarterly Bulletin; Marine Corps Gazette*
JP 3-0 (*Doctrine for Joint Operations*, 1993), 137, 138, *138*
JP 3-07 (*Military Operations other than War*, 1995), *138*, 142–44, 206
JP 3-07.3 (*Tactics, Techniques, and Procedures for Peace Operations*, 1999), *138*, 144
JP 3-08 (*Interagency Operations*) vols. 1 and 2: on CORDS as example of civil-military coordination, 167; on expectations for civilian agencies, 137–38; with lessons from previous operations, *138;* MOOTW included in, 144; on responsibility vs. capacity or capability of civilian agencies, 169–70
judge advocate general's (JAG) Corps, 58
Justice, U.S. Department of, 166
just-in-time learning, Center for Army Lessons Learned and, 105

Kagan, Fred, 176–77
Kahl, Colin, 181
Keane, Jack, 176–77
Kier, Elizabeth, 14, 18
Kilcullen, David, 178–79, 182–83
kinetic actions, 205
King, Alan, 165
Kitfield, James, 111
Kolb, David A., 110
Komer, Robert, 75–76
Korean Peninsula, Bottom-up Review (1993) on potential challenge to U.S. in, 85
Kosovo: JRTC training for, 123; post–Cold War length of mission in, 72. *See also* Balkans
Krepinevich, Andrew F., Jr., 74–75, 175–76
Krulak, Charles C., 146–47, 185
Kuwait, Iraq's strained relations with before U.S. invasion of Iraq, 162

laws of war, as resources for Philippine and Cuban campaigns, 39–40, *39*
leadership, role in organizational learning theory, 26
learning cultures, in civilian agencies, stabilizing complex societies and, 161
learning cycle: in organizational learning theory, 22–23, *23;* professional military education and, 156–57. *See also* military learning
learning system: early civil-military conflict in Iraq and, 160–61; four-block war and limits of, 185–88; military's capacity for institutional learning and, 199–200; military's interagency theorizing for Iraq and, 172
learning through doing: combat training centers and, 195–96; promoted at National Training Center, 112
Learning to Eat Soup with a Knife (Nagl), 160
Lebanon, Cold War peacekeeping operations in, 70

legal issues, institutionalizing civil affairs and military government and, 58
Lejeune, John A., 52
Lend-Lease Administration, Allied occupation in North Africa and, 63
Leonard, Steve, 153, 172–73
lessons-learned system: formalized in 1990s, 195; four-block war learning system and, 190; institutional amnesia on MOOTW and, 193; officer commissioning class of 1995 and, 202; post–World War I, 54–56; U.S. Army's success in Iraq due to use of, 189, 200. See also doctrine; military learning
Levy, Jack S., 24
Lieber Code, 33, 39, 57. See also General Order 100
Light Infantry: combat training centers and development of, 99; JRTC and, 116; JRTC development and, 113; reorganization of 2005 and, 116n; White Paper on operations of, 114–15
Lincoln, Abraham, 33
low-intensity conflict (LIC), 135, 205–6. See also counterinsurgency; military operations other than war; stability operations
Lute, Douglas, 171

MacKenzie, Lewis, 81
Magruder, Lawson W., III, 120
Mahan, Dennis Hart, 31
major war. See war, big or major
al-Maliki, Nuri, 185
Mansoor, Peter, 178–79
March to the Sea, Sherman's, 33–34, 40
Marine Corps, U.S.: Army on infrastructure reconstruction vs., 51–52; cultural disposition to informally share knowledge in, 20–21; cultural education program on MOOTW of, 199; hierarchy of schools for professional military education, 154–55; history of MOOTW experiences by, 10, 191; incorporating MOOTW into doctrinal manuals and, 151; land operations prior to Banana Wars for, 47; learning system evolution during Banana Wars for, 45, 46; political resistance to learning among, 52–53; Root and Stimson on mission of Army vs., 44; in Somalia, Vietnam syndrome and, 79–80; taking control of Nasiriya, Iraq, 1; use of "irregular warfare" as umbrella term by, 205; use of "small wars" by, 207. See also Small Wars Manual; U.S. military
Marine Corps Gazette, 48, 49–50, 194
Marshall, George C., 60, 64
Mattis, James, 20, 132, 152, 177–79, 198
McMaster, H. R., 174, 178–79
Meese, Mike, 178–79
Mentis, Alex, 109–10, 109n
METLs (mission essential task lists), 99, 119, 141–42, 142n
Meyer, Edward C. "Shy," 98, 99, 114, 115
Middle East, Bottom-up Review (1993) on potential challenge to U.S. in, 85
MILES (Multiple Integrated Laser Engagement System) gear, 101, 117
Miles, Nelson A., 36
Military Aid to the Civil Power (1925), 57–58, 59
military change: bureaucratic politics and, 13–14; Clausewitz on military learning and, 9; doctrine and education and, 129–30; integrated theories of, 16–19; through learning, internal structures and processes and, 192–93; modern theories of, 10–11; organizational culture and, 14–15; as organizational learning, 19–22; organizational theory on, 11–13; professional military education as engine of, 154–57; role of leadership in, 26; top-down, using doctrine, 131. See also bottom-up learning; military learning; top-down systems
military duties, posse comitatus separating police duties from, 35–36
Military Force Structure Review Act (1996), 86n
Military Government, 59
military government: debate on soldiers' roles in civil affairs and, 61–64; differences between civil affairs and, 57n; institutionalizing civil affairs and, 56–59
Military Government and Civil Affairs (FM 27-5, 1940), 59, 59, 60
Military Government and Martial Law (Birkhimer, 1892), 39
Military Government Division memo, on civil administration of military governments, 62
Military Institutes (official U.S. military regulations), 31
Military Law, 58, 59
military learning: changing role of doc-

trine and education in cycle of, 129–32; civil-military debate over soldier's roles, 61–64; Clausewitz on military change and, 9; combat training centers and, 99–100; communities of practice and other 21st-century informal networks, 124–26; coping and adapting in early years of U.S., 28–31; coping and adapting in the Philippines and Cuba, 38–41; development of *posse comitatus* and, 34–36; evolution of modern system of, 98–102; experiential to organizational, CTC model and, 110–12; frustrations identified in World War I, 53–56; institutionalizing civil affairs and military government, 56–59; institutional learning, political resistance and, 65; Joint Readiness Training Center, 112–18; JRTC original focus on MOOTW and, 118–23; lessons from the Philippines applied to Cuba, 42–45; Marine Corps records learning during Banana Wars and, 45; Marine Corps tactics, techniques, and politics in Banana Wars, 46–50; National Training Center, 100–102; political resistance to, 52–53; post–Vietnam War institutional changes and, 3; post–Vietnam War training programs, 97–98; the School of Military Government, 59–61; *Small Wars Manual* and, 50–52; U.S. Army's organizational processes and policies in Vietnam and, 75–76; during U.S. Civil War and Reconstruction, 32–34; during Vietnam War, 76–77; Wagner's contributions to role of, 36–38; what has worked and how, 194–95. *See also* bottom-up learning; Center for Army Lessons Learned; learning cycle; military change; top-down systems

military operations, post–Cold War qualitative changes and trends in, 70–73

military operations against irregular forces, 138

Military Operations in a Low-Intensity Conflict (FM 100-20), 122–23, 134, *135,* 136

military operations other than war (MOOTW): big-war paradigm of U.S. Army and, 73–74; Commission on Roles and Missions on, 86–87; conducted by U.S. military by 1996, 89; de-emphasized in FM 3-0 (2008), 145; doctrine on, 132–34, 134–38, 189–90, 197–98;

failure to adjust doctrine and education to support, 129; FM 100-5 chapter on, 139–40; GAO reports (1995–1999) on U.S. military readiness and, 89–90; history of Army and Marine Corps' experience with, 10; institutional amnesia vs. lessons learned on, 193; next generation of doctrine for, *138;* operations manuals (1993) include chapters on, 138–39; organizational culture and, 15; original focus of JRTC on, 118–23; permitted but not promoted in professional military education, 199; post–Cold War bureaucratic politics model and, 13–14; Powell on U.S. military role in, 82–83; PPD-25 on, 84–85; in Somalia, Vietnam syndrome and, 79–80; use of term, 206; U.S. military history with, 191. *See also* peacekeeping operations; stability operations

military police, Gullion's school in World War II for, 60

Military Review, 121–22, 183, 188–89

Military Support for Stability, Security, Transition, and Reconstruction (SSTR) Operations (DoDD 3000.05), 207

Miller, Jesse, 62

Millett, Allan R., 53

mission, U.S. military: Bottom-up Review (1993) of, 85. *See also* METLs; nonmilitary missions

mission creep, Weinberger-Powell Doctrine on, 78

mission essential task lists (METLs), 99, 119, 141–42, 142n

mission rehearsal exercises (MREs), 196, 201. *See also* training

Modular Force, U.S. Army as, 109–10

MOOTW. *See* military operations other than war

multilateral forces: post–Cold War operations and, 70–71; U.S. commitment to UN on, 82

Multi-National Force–Iraq (MNF–I), 173, 182–84

Multiple Integrated Laser Engagement System (MILES) gear, 101, 117

Multi-service Manual for TTP for Peace Operations (FM 3-07.31, 2003), 144, 151

Murphy, Robert, 63–64

muscle memory development: combat training centers and, 99; flight training programs and, 100; JRTC and, 113, 117

Nagl, John A.: on bottom-up organizational learning model, 160; FM 3-24 (*Counterinsurgency*) development and, 152–53; on organizational culture, 15; on organizational learning by U.S. Army vs. British Army, 20, 192; on U.S. Army's learning system in Vietnam, 76, 188–89
Nagle, Bill, 183
Napoleonic Wars, 29, 36
Nash, William "Bill," 88, 92–93
National Defense Act (1920), 57, 65
National Defense Authorization Act (FY1994), Commission on Roles and Missions and, 86
National Defense University, 170
National Military Strategy (NMS), 86, 86n, 87–88
National Security Strategy (NSS), on 2-MTW, 85–86
National Training Center (NTC), Fort Irwin, Calif.: adaptations for JRTC experience at, 123; criticism of mission of, 113–14; cultural paradigm of military to fight big wars and, 97–98; education and training on AirLand Battle Doctrine at, 195; focus of JRTC vs., 116; GAO report on Center for Army Lessons Learned and, 103; JRTC development and, 113; mission and characteristics of, 100–102; officer commissioning class of 1975 training at, 200; precursors to, 99–100; promoting learning through doing at, 112
nation building: American political distaste for, 91–92; in Cuba after Spanish-American War, 42–43; institutional processes hindering teaching lessons learned from, 65; as original purpose for U.S. Army, 28–29; PPD-25 on, 84–85; stability ops compared to, 4; after U.S. Army's 19th-century campaign against Indians and, 30–31; Vietmalia syndrome as aversion to, 90–91
Navy, U.S., pilot training by, 99–100. *See also* U.S. military
Newbold, Gregory, 79, 91, 91n, 137
News from the Front! 108
new world order: doctrine development for, 138–47, *138;* peace dividends vs., 68–70; peacekeeping operations in Somalia, Vietnam syndrome and, 79–85; qualitative changes and trends in military operations, 70–73; real mission, 2-MTW, and readiness debate, 85–90; U.S. military and, 67–68; Vietmalia syndrome from Haiti to Iraq, 90–95
New York Civil Liberties Union, 92
Nicaragua: lessons from Haiti applied to Marine Corps in, 47–48; Marine Corps's pacification duties in (1914–1934), 45; Marine Corps tactics, techniques, and politics in, 46–50; political constraints on Marines in, 51
NMS (National Military Strategy), 86, 86n, 87–88
nongovernmental organizations (NGOs): cited in FM 100-23 (*Peace Operations*), 141; cited in JP 3-07 (*MOOTW*), 143–44; Emerald Express (civil-military exercise) and, 168–69; JP 3-08 on responsibility and capability of, 169–70; JRTC training and, 121; Marines and Army in Somalia operation reach out to, 168; post–Cold War operations and, 70, 71; vetting language of FM 3-07, 172–73
nonmilitary actors, post–Cold War operations and, 70, 72. *See also* civilian agencies; external actors or pressure
nonmilitary missions, terms used to describe, 3–5, *4*
Noriega, Manuel, 67
North Africa, Allied invasion of, 63–64
Notes on Recent Operations (AEF, World War I), 55–56
NSS (National Security Strategy), on 2-MTW, 85–86
NTC. *See* National Training Center
Nunn-Cohen Amendment (1987), 115

Oakley, Robert, 121
observer controllers (OCs), 101, 106, 111–12, 117
occupation: doctrinal manuals on duties during, 58; Marine Corps' learning during Banana Wars about, 45; studying Army's role in, 195–96; by U.S. Army, Reconstruction as, 34–36; U.S. Army program of duties for, 53. *See also* Cuba; military operations other than war; Philippines
Odierno, Raymond, 179, 183, 184
offensive doctrines, debate on defensive doctrines vs., 18n
"Old and New Testaments of American Military Government," 59
Oliver, George, 147, 151

On Strategy: The Vietnam War in Context (Summers), 74, 78n
On War (Clausewitz, 1831), 29
Operation Desert Storm, 130, 139. *See also* Iraq
Operation Just Cause, 67, 131. *See also* Panama
Operation Provide Comfort, 131, 141. *See also* Iraq
Operation Provide Hope. *See* Somalia
Operation Restore Hope, 137, 141. *See also* Somalia
Operations. See FM 3-0; FM 100-5
operations analysis, Vietnam War and, 76–77
Operations in a Low-Intensity Conflict (FM 7-98), 123
operations manuals. *See under* FM; *under* JP
operations other than war (OOTW): in FM 100-5, 139. *See also* military operations other than war
Operations other than War, volume 4, *Peace Operations*, 123
Operation TORCH (Allied invasion of North Africa), 63–64
Operation Uphold Democracy, 71, 143. *See also* Haiti
OPFOR (opposing force): for JRTC training, 117–18, 121; for National Training Center, 101
organizational culture: CTC learning environment model and, 111–12; military change and, 14–15; Nagl on organizational culture and institutional learning, 192; Rosen on military change by maverick officers and, 17; U.S. Army's success in Iraq due to shift in, 189–90. *See also* cultural resistance to change
organizational learning: as foundation for the Surge, 182–85; institutional learning phases in, 175; military change as, 19–22; theory of, 22–26. *See also* experiential learning; military change
organizational personality, Builder on, 14
organizational theory, on military learning and military change, 11–13
Organization and Tactics (Wagner, 1901), 39, 43
O'Sullivan, Meghan, 176
Outlines of International Law (Davis, 1888), 39
outside pressure. *See* external actors or pressure

Pace, Peter, 174
Panama: adequacy of LIC manual on challenges in, 136–37; doctrinal support for Operation Just Cause in, 134; doctrine-writing after operational lessons learned in, 133; Operation Just Cause in, 67; post–Cold War length of mission in, 72
paradigms, challenges to, in institutional learning, 175
Parker, James, 39
peace dividends, new world order vs., 68–70
Peacekeeping and Stability Operations Institute (PKSOI), at Army War College, 150–51
Peacekeeping Institute (PKI), at Army War College, 119, 148–52, 168, 189, 195
peacekeeping operations: Clinton's plan for, 80–81; Commission on Roles and Missions on, 86–87; debate on U.S. military readiness and, 88–89, 88n; doctrinal writing in the 1990s for MOOTW and, 132–34; FM 100-23 (*Peace Operations*, 1994) and training in, 142; full-spectrum operations and, 144–47; JRTC training for, 120; mission objectives, murky political guidelines and, 121–22; during 1990s, conflicting mandates for, 90–91; planning failures for Iraq during postwar period, 163–66; post–Cold War, 70–71; PPD-25 on, 83–85; QDR and NMS of 1997 on, 87–88; resisted by leadership in favor of AirLand Battle Doctrine, 157; in Somalia, Vietnam syndrome and, 79–80; by U.S. Army of Indians during 19th century, 30; U.S. National Security Strategy on 2-MTW and, 85–86. *See also* military operations other than war
Peace Operations (FM 100-23, 1994), *138*, 140–41
Pentagon: bottom-up learning in Iraq complicated by, 190; *Military Support for Stability, Security, Transition, and Reconstruction (SSTR) Operations* (DoDD 3000.05) of, 207; planning, programming, budgeting, and execution (PPBE) cycle, 11–12; post-1989 strategic and political debates in, 67–68. *See also* Defense, U.S. Department of
Pershing, John, 54–55

Persian Gulf War: Cheney on not ousting Saddam Hussein during, 162; length of, 72; U.S. plans for downsizing military during, 68–69
Peru, post–Cold War peacekeeping operations in, 70
Petraeus, David: Alywin-Foster on success in Iraq by, 189; exploitation of Sunni Tribal Awakening by, 181–82; on force increases for Surge, 174; Odierno on Surge in Iraq and, 184; as officer commissioned in 1975, 201; operations in Iraq and Afghanistan and doctrinal development under, 152; rewriting counterinsurgency field manual, 177–79; stability and reconstruction plan for Iraq of, 94–95; 21st-century revisions to Army doctrine and, 132; use of Surge and new learning culture to stabilize Iraq, 7, 159–60
Philippine Constabulary, 41
Philippine Insurrection against the United States, The (Taylor), 44
Philippine Macabebe Scouts, 41
Philippines: coping and adapting after Spanish-American War in, 38–41; Marine Corps' view of Army operations in, 47
planning, programming, budgeting, and execution (PPBE) cycle, Pentagon's, 11–12
PlatoonLeader.com, 124, 183
police duties, *posse comitatus* separating military duties from, 35–36
political systems: Avant on military change and, 17–18; as dampener on military learning for nontraditional military missions, 45; as institutional processes hindering teaching lessons learned from controversial roles, 65. *See also* external actors or pressure
Posen, Barry R., 16
Posse Comitatus Act (U.S., 1878), 35
post–Cold War bureaucratic politics model: doctrine-writing and, 133–34; military operations other than war and, 13–14; peace dividends and new world order of, 68–70. *See also* Cold War; new world order
Powell, Colin, 69, 78, 82–83
Powell Doctrine, 78–79, 84–85
president, U.S.: annual National Security Strategy of, 86n; on soldiers' roles in military governments, 61–62. *See also specific presidents*
Presidential Decision Directive-13 (PDD-13), 81–82, 81n
Presidential Decision Directive-25 (PDD-25), 82, 85, 141
Presidential Review Directive 13 (PRD-13), 81–82, 81n
probability theory, Vietnam War and, 76–77
professional military education (PME): military change and, 154–57; as weak link in post–Cold War learning system, 198–200
progressivism, Cuba as workshop for, 42–43
provincial reconstruction teams (PRTs), for Iraq, 171–72
Prussian military style, American interest in, 37–38
public works, U.S. Army vs. Marine Corps on reconstruction of, 51–52. *See also* infrastructure

Quadrennial Defense Review, 12, 86, 86n, 87–88
Quantico, Va.: lectures on lessons learned in Haiti by 1927 at, 47–48; Lejeune's promotion of schools at, 52; *Small Wars Manual* written at, 45, 194. *See also* Command and Staff College

RAND Corporation, 136–37
Rangers (Army light battalion), 116
readiness, debate over definition of, 85–90
reconstruction. *See* stability and reconstruction operations
Reconstruction (post–Civil War): U.S. Army and rebel insurgents during, 32–34; as U.S. Army occupation, 34–36
Red Flag flight training program, U.S. Air Force's, 100
refugees: in JRTC "battlefield," 121; post–Cold War operations and assistance for, 71. *See also* infrastructure
Remarks concerning Deficiencies in the Training of Our Units as Brought Out in Some of the Recent Offensive Operations (AEF, World War I), 55–56
Report of American Officers on Recent Fighting (AEF, World War I), 55–56
reports, battle or field: early uses for, 31; Wagner's review of, 37. *See also* after-

action reviews; journal articles; trend reports
request for information (RFI) page, of CALL website, 109
reserve units, for long-term missions, 72–73
Rice, Condoleezza, 171–72
Ricks, Tom, 27, 165, 184
Rockwood, Lawrence, 92
Roosevelt, Theodore, 42
Root, Elihu, 44
Rosen, Stephen P., 16–17
rotation systems, personnel: individual, during Vietnam War, 77; of units, for long-term missions, 72–73
rule of law: as concern of U.S. military, 2; in Haiti, U.S. military concerns about, 92; three-block war and, 186. *See also* infrastructure
Rules of Land Warfare, The (FM 27-10, 1939), 58, 59, *59*, 60
Rumsfeld, Donald, 164
Russell, John, 52–53

Saudi Arabia, Iraq's strained relations with before U.S. invasion of Iraq, 162
scenarios. *See* training
School of Advanced Military Studies (SAMS), Army's, 155–56, 195
School of Advanced Warfare (SAW), Marine's, 155–56
School of Military Government, at University of Virginia, 59–61, *59*
schools, military: hierarchy of, 154–55; on-site in Haiti and the Dominican Republic (1915–1934), 48–49. *See also* Fort Leavenworth, Kans.; Quantico, Va.; U.S. Military Academy
Scott, Winfield, 31
Second Armored Cavalry Regiment's Third Squadron, 122
security forces in post-invasion Iraq, U.S. military planners on, 165. *See also* infrastructure
Seminole Indians, U.S. Army's campaign (1817–1842) against, 30
Senge, Peter M., 20, 21, 23
Sentiments on a Peace Establishment (Washington, 1783), 29
Service of Security and Information, The (Wagner, 1896, 1899), *39*, 43
Sewall, Sarah, 178
Shalikashvili, John, 69, 83, 122, 143, 143n

Sherman, William Tecumseh, 32–34, 41
Shia majority, in Iraq, 162
Shinseki, Eric K., 108, 144–45, 164
Shugart, Randall, 80n, 118
Sinai peninsula, Cold War peacekeeping operations in, 70
small wars: AirLand Battle Doctrine and, 114; diplomatic pressures during, 46–47; leadership for Banana Wars on amphibious warfare vs., 157; Marine Corps adds lectures at Quantico on, 48; Marine Corps emphasis on amphibious warfare vs., 52–53; use of term, 207; U.S. military learning and, 27–28. *See also* military operations other than war
Small Wars and Punitive Expeditions (19th-century Army book), 50
Small Wars Journal blog/website, 183
Small Wars Manual: Banana Wars (1914–1934) and development of, 45, 207; as formal adaptation of military learning, 28; on major war vs. small war, 46–47; Marine Corps' writing and publication of, 50–52; Marines' rerelease in 1987 of, 135; precursors to, 49–50; reflecting lessons from other militaries, 10; transfer of Banana Wars experience to, 194; use in Marine officer training, 1
Small Wars: Their Principles and Practice (Callwell, 1896), 49, 207
Somalia: adequacy of FM 100-5 and JP 3-0 for challenges in, 137; adequacy of LIC manual on challenges in, 136; doctrinal support for Operation Provide Hope in, 134; doctrine-writing after operational lessons learned in, 133; learning through doing in, 195; lessons learned from, 126; light forces deployed to Haiti after deployment to, 116–17; military attempts at coordinating with civilian agencies in, 168; as model for JRTC training, 118; nebulous orders for U.S. Marines' 1992 peacekeeping operation in, 91; PDD-25 and, 84; PKI development and Army's experience in, 148–49; post–Cold War peacekeeping operations in, 70; Ricks on U.S. peacemaking (1992) in, 27; troop types used in, 71–72; U.S. military and new world order in, 67–68; U.S. military participation in peacekeeping and tragedy in, 83; U.S.

Somalia (*continued*)
role in UN peacekeeping and tragedy in, 82; veterans of, on PME curricula relevance after, 156; Vietnam syndrome and peace operations in, 79–80

Sources of Military Doctrine, The (Posen), 16

Soviet Union: flight training for conflict with, 100; implosion as primary enemy, 13, 13n; light divisions for fight against, 115; new world order and, 68; NTC and readiness for conflict with, 112; as primary enemy, 67, 77; rebuilding Hollow Force against, 99, 114; slow development of JRTC and threat by, 116; training for conflict with, 72

Special Forces, 99, 116, 116n, 189–90

Special Operations Command (SOCOM), 135

Special Operations Forces, 113, 115, 201, 205

stability and reconstruction operations (S&R): capacity and capability issues for, 166–67; complexity in Iraq, scale and strategic nature of, 161; definition of, 3–4; evaluating success of, 186–87; history of U.S. military's engagement in, 9; *posse comitatus* and, 35–36; after Spanish-American War, 36; U.S. military learning system and, 6–7; U.S. military preparation for, 1–2; U.S. military's study of, 10. *See also* irregular warfare; stability operations

Stability and Support Operations. See FM 3-07; FM 100-20

Stability and Support Operations Handbook, 108

stability operations (stability ops): civilian actors in lessons-learned system for, 188; definition of, 3, 207; doctrinal manuals in late 1990s on, 134; incorporated into Army's PMW curriculum, 199; Light Infantry and Special Forces for, 116; next generation of doctrine for, *138;* PMA relevance for institutional learning on, 156–57. *See also* stability and reconstruction operations

State, U.S. Department of: capacity and capability issues for S&R of, 171–72; on counterinsurgency, 203; on insurgency, 204; lack of expeditionary capacity of, 169; military assumption on deployable experts to Iraq from, 166; Office of Foreign Relief and Rehabilitation, Allied occupation in North Africa and, 63; support for operational capacity of UN peacekeeping operations and, 82

Steiner, Carl, 140

Stimson, Henry, 44

St. John's College, Cambridge, England, 59, 60

Strategy for Victory in Iraq (White House, 2007), 174

Stroup, Theodore G., Jr., 14–15

Stryker Brigade Combat Team, 116

subject matter experts (SMEs), 103, 105, 106

Sullivan, Gordon: on Army leadership and organizational learning, 21; on CompanyCommand.com, 124n; on CTC program and organizational learning, 112; debate over doctrine developed by, 131–32; on interacting with NGOs, 121; MOOTW chapter in FM 100-5 and, 139, 140; on PKI role for U.S. Army, 148–49, 151–52; on post–Cold War process of writing new doctrine, 131; systems for generating knowledge and capturing lessons and, 133; training for MOOTW and, 119

Summers, Harry H., Jr., 74, 78n

Sunni minority in Iraq, 162, 180–82

Surge for Iraq (adopted 2007): advocacy and intellectual foundations of, 175–77; assessing, 180–82; as change in tactics and longer-term approach, 173; doctrinal foundation for implementation of, 177–80; four-block war challenge and, 185–88; military learning and success of, 7; organizational learning as foundation for, 182–85

Swannack, Charles H., 120, 122–23

SWETI (sewage, water, education, trash, and information) lines of operation, 189

"Synopsis of Military Government" (1942), 62–63

Syria, Iraq's strained relations with before U.S. invasion of Iraq, 162

Tactics (FM 3-90), 147

take-home packages (THPs), 101, 117

Task Force Hawk, Kosovo, 106

Taylor, J. R. M., 44

Teller Amendment (1898), 42

terrorists, JRTC training for, 118

Thomson, Charles, 60
three-block war, 146–47, 185, 186
Thurman, Max, 99
top-down systems: Army communities of practice in, 124–26; budgets and, 12; doctrine as tool for change in, 131–32; doctrine-driven organizations as, 6; education and training systems as doctrinal change tool in, 155–56; generational split between writers and sponsors of doctrine in, 157; Pershing's, facilitating bottom-up learning in, 54; political resistance to learning in, 65; post–Vietnam learning system as, 195–96, *196;* unrealistic expectations of nonmilitary government agencies with, 160–61; U.S. military theories written into doctrine in, 172
Top Gun flight training program, U.S. Navy's, 100
torture, water cure used in Philippines War, 41, 41n
TRADOC (Training and Doctrine Command) proponent schools, 105, 106, 151
training: doctrine, education, and the revolution in, 195–200; four-block war learning system and scenarios for, 186; Joint Readiness Training Center scenarios for, 116–18
Treasury, U.S. Department of, 166
trend reports, CTC, 106
Tribal Awakening, among Sunni tribes, 180–82
Troops in Campaign: Regulations for the Army of the United States (1892), 39, 40
Troops in Campaign: Regulations for the Army of the United States (1903), 43
troop types, Cold War vs. post–Cold War, 71–72
25th Infantry Division (Light), Second Brigade, JRTC training for, 120
two major theater war (2-MTW) strategy, 85, 87–88

unconventional warfare (UW), 208
unconventional warfare operations, 138
United Nations, 81–82, 141
United Nations Operation in Somalia I and II, 80
University of Chicago Press, 153, 178
University of Virginia, School of Military Government at, 59–61, *59*

Upton, Emory, 73
Urban Operations (FM 90-10, 1979), 134, *135*
urban operations, after-action reviews on Operation Just Cause in Panama and, 136–37
U.S. Agency for International Development (USAID), 166, 167, 168
U.S. Information Agency, 167
U.S. Institute of Peace, 175, 177
U.S. military: Bottom-up Review (1993), 85; culture of, Iraq planning and, 163–66; history of MOOTW experiences by, 191; as "Hollow Force," Meyer on, 98; MOOTW conducted by 1996 by, 89; post–Cold War downsizing of, 69; Powell on role in MOOTW-type missions for, 82–83; on responsibility vs. capacity or capability, 169; use of "irregular warfare" as umbrella term by, 205. *See also* Air Force, U.S.; Army, U.S.; Marine Corps, U.S.; Navy, U.S.
U.S. Military Academy, 29, 43. *See also* CompanyCommand.com/Company Command.mil
"U.S. Policy on Reforming Multilateral Peace Operations" (PDD-25), 82
Utley, Harold H., 49, 50

Vetock, Dennis J., 37, 38, 54, 55, 76–77
Vietmalia syndrome, 84–85, 90, 119
Vietnam syndrome, military operations other than war and, 79–85
Vietnam War: Air-Land Battle Doctrine and change after, 130; AirLand Battle Doctrine dissemination after, 151–52; American Army resistance to civilian intervention during, 17–18; Civilian Operations Revolutionary Development Support (CORDS) in, 167–68; Cold War attitudes and lessons from, 73–77; counterinsurgency warfare in, 53; generational learning after, 24–25; U.S. Army's learning culture after, 21; Weinberger-Powell Doctrine and, 78–79
volunteers: long-term missions as strain on, 73; U.S. Army comprised of, 97–98

Wagner, Arthur L., 36–38, 50
Waller, Littleton, 42, 46
Walters, Steve, 109n

war, big or major: diplomatic relations at conclusion of, 46–47; guidance for MOOTW or peace operations and readiness for, 133; lessons learned vs. cultural disposition for, 193; QDR and NMS of 1997 on, 88–89; as U.S. Army paradigm after Vietnam War, 97–98; as U.S. Army paradigm from World War I to Vietnam, 73–75; U.S. Army's Iraq invasion and predisposition to, 188; U.S. military preparation for stability operations or counterinsurgency vs., 1–2. *See also* drug war; irregular warfare; small wars; two major theater war strategy

war colleges: resistance to change educational curricula of, 97–98; senior level professional military education at, 154–55. *See also* schools, military

War Department, U.S., 28, 35, 64

War Powers Act, Clinton's use of, 80

Washington, George, 29

websites: Center for Lessons Learned, 108–9; for communities of practice, 124, 125–26, 125n, 183; exposure to MOOTW through, 199; four-block war learning system and, 186; informal networks through, 198; military learning through, 97

Weigley, Russell F., 29, 57

Weinberger, Caspar, 78

Weinberger Doctrine, 77, 78–79

Wenger, Etienne, 124–25

West Point, counterterrorism site of, 183. *See also* U.S. Military Academy

Wheaton, Loyd, 41

White House. *See* president, U.S.

Wickersham, Cornelius, 59, 60–61

Wickham, John A., 114–15

Wolfowitz, Paul, 164

Wood, Leonard, 42–43

Wool, John E., 34

worldviews, generational change and, 24–25

World War I, AEF's G5 training section during, 53–56

World War II: Allied invasion of North Africa, 63–64; American Expeditionary Force in, 54, 55–56; amphibious warfare focus of Marines during, 52–53, 194; Gullion's school for military police during, 60

Worth, William J., 30

Ya'alon, Moshe, 108

Yugoslavia, former. *See* Balkans

zero-defect mentality, military tolerance for mistakes and, 17n31

Ziemke, Earl F., 57–58

Zinni, Anthony, 133, 168–69

Zisk, Kimberly M., 18

Printed and bound by CPI Group (UK) Ltd, Croydon, CR0 4YY
09/06/2025

14685670-0002